TRANSPORT IN SKELETAL MUSCLE

TRANSPORT IN THE LIFE SCIENCES

EDITOR: E. EDWARD BITTAR

VOLUME 1 *Cell Potassium* — Roderick P. Kernan

VOLUME 2 *The Sarcoplasmic Reticulum: Transport and Energy Transduction* — Leopoldo de Meis

VOLUME 3 *Transport in Skeletal Muscle* — Raymond A. Sjodin

TRANSPORT IN SKELETAL MUSCLE

Raymond A. Sjodin
Department of Biophysics
University of Maryland
School of Medicine

1807 175 YEARS OF PUBLISHING 1982

A Wiley-Interscience Publication
JOHN WILEY & SONS
New York • Chichester • Brisbane • Toronto • Singapore

Library of Congress Cataloging in Publication Data:

Sjodin, Raymond A.

Transport in skeletal muscle.

(Transport in the life sciences, ISSN 0271-6208; v. 3)

"A Wiley-Interscience publication."

Includes index.

1. Striated muscle. 2. Ions—Metabolism. 3. Biological transport, Active. 4. Ionic mobility. I. Title. II. Series. [DNLM: 1. Ions—Metabolism. 2. Muscles—Metabolism. W1 TR235V v.3 / WE 500 S6256i]

QP321.S493 591.1′852 82-3446
ISBN 0-471-05265-5 AACR2

Printed in the United States of America

10 9 8 7 6 5 4 3 2 1

To Marion,
Lisa, and Neva

SERIES PREFACE

Membrane transport is rapidly becoming one of the best-worked fields of modern cell biology. The Transport in the Life Sciences series deals with this broad subject in monograph form. Each monograph seeks to trace the origin and development of ideas in the subject in such a way as to show its true relation to membrane function. It also seeks to present an up-to-date and readable outline of the main problems in the subject and to guide thought on to new lines of investigation.

The task of writing a monograph is not a light one. My special gratitude is to the various authors for expounding their subjects with scholarly care and force. For the preparation of the indexes I thank Dr. Barbara Littlewood.

E. EDWARD BITTAR

Madison, Wisconsin
June 1980

PREFACE

This monograph could have been written from a number of different approaches, but I have chosen a historical approach, using the material most familiar to me and drawing heavily on my own laboratory observations. I feel that this orientation imparts more first-hand authenticity to the manuscript and produces a more readable work. This would not have been possible, in my opinion, had I taken a more comprehensive tack.

In essence, I have elected to develop the ideas underlying transport in skeletal muscle as they have evolved in a single muscle type, namely frog muscle. Some results with other kinds of muscle have been cited at intervals to help underscore similarities and differences. Frog skeletal muscle was chosen as a general model for muscle transport for several reasons. Research on ion distribution, ion transport, and the membrane potential had early beginnings in frog skeletal muscle. From these beginnings, a steady accumulation of kinetic information on ion transport has occurred. The first convincing demonstration of the sodium pump's contribution to the resting potential of excitable cells was done on frog muscle. Transport studies in other skeletal muscle types often involve kinetic problems in flux measurement that are difficult to surmount. The study of transport in excised mammalian muscle, for example, is made difficult by extracellular diffusion problems in whole muscle and by the greater dependence on an intact circulatory system. The fact that the frog has a thin, flat, and easily isolatable sartorius muscle that survives well without intact circulation has helped to solve many problems.

The monograph begins with some first principles and some basic early work and proceeds with the evolution of the ion transport studies that have placed skeletal muscle within the same framework as red blood cells and giant axons. An attempt is made to say something about all of the major components of ionic flux that are believed to occur in skeletal muscle fibers. There is a certain risk in doing this, as little is known about some components of ionic flux in muscle. Indeed, some observations are confusing and no suitable explanation yet exists for them. A more "airtight" picture could have been developed for ion transport in muscle by simply deleting the confusing parts. I feel, however, that the best interests of the reader would not be served by this procedure. The reader is thus invited to participate in the process of deciding which aspects of transport in muscle are sensible and which must await further research for clarification.

The monograph is greatly enhanced by the illustrations, a number of which have been reproduced from the literature. I thank all the authors, the *Journal of General Physiology*, the *Journal of Physiology*, and Cambridge University Press for permission to reproduce figures. I am also grateful to Professor E. Edward Bittar, Editor of this series, for giving me the opportunity to do this work. Special appreciation goes to Dr. Lorin J. Mullins for reading the manuscript and to my wife, Marion, for editorial assistance. Finally, I thank Ms. Esther Scheffel and Ms. Kathy Artwohl for their very capable preparation of the typed manuscript.

RAYMOND A. SJODIN

Baltimore, Maryland
April 1982

CONTENTS

TRANSPORT IN SKELETAL MUSCLE

1

Conceptual Framework for Ion Movements in Skeletal Muscle

It is easy to overlook the fact that skeletal muscle transport studies began long before the electron microscope enabled the more dramatic observations on muscle contraction. Yet as early as 1934, Fenn and Cobb had published the results of their investigation of the K distribution in frog sartorius muscle.

Though pertinent earlier work does exist, it is fitting to set the beginning of muscle transport research with the K studies of Fenn and Cobb.[1] Potassium is *the* major cation in muscle fibers—it contributes about half their osmotic pressure. Potassium distribution is also the major determinant of the muscle membrane potential; this is another reason why the K studies are a logical place to begin.

The basic finding of Fenn and Cobb was that the tendency of muscle fibers to gain or lose K ions is determined by the extracellular K ion concentration, $[K]_o$. Sartorius muscles from the frog neither gained nor lost K ions over a period of many hours if $[K]_o$ was maintained at a value of 4.9 mM. However, muscles lost K if $[K]_o$ was lower than 4.9 mM and gained K if $[K]_o$ exceeded this value. As intracellular K ion concentration in frog sartorius muscle fibers is about 140 mM,[2,3] flux balance for K ions is achieved at a $[K]_i/[K]_o$ ratio of 28.6.

TYPES OF ION DISTRIBUTION AND MOVEMENT

Electrodiffusional Equilibrium

For an ion, j, in electrochemical equilibrium, the equilibrium potential, E_j, for that ion is given by

$$E_j = \frac{RT}{zF} \ln \frac{[C_j]_o}{[C_j]_i} \tag{1}$$

where R is the gas constant, z the ionic valence, F the Faraday constant, T the absolute temperature, and $[C_j]_o$ and $[C_j]_i$ the external and internal concentrations of the ion of species j, respectively. Under equilibrium conditions, an internal electrical negativity of -84.5 mV inside the muscle fibers would account for the steady state of the muscle K. As the muscle membrane potential approximates this value when $[K]_o = 5$ mM,[4,5] the muscle K is close to meeting the conditions for electrochemical equilibrium. At lower values of $[K]_o$ and at high values of $[Na]_i$, departures from equilibrium become evident. Such departures suggest the operation of additional forces, which are discussed in subsequent chapters.

The intracellular chloride concentration is low in muscle fibers, between 3.1 and 3.8 mM in frog sartorius muscles when $[K]_o = 2.5$ mM and $[Cl]_o =$ 120 mM.[6] Application of equation (1) to the Cl concentrations under these conditions yields an electrical potential that is in good agreement with measured values of the muscle fiber resting membrane potential.

Measurements of transmembrane fluxes of K and Cl, using radioactive tracers, indicate appreciable membrane permeability to these ions in muscle.[6-9] These results are consistent with the interpretation that KCl in muscle is close to a Donnan distribution, in which KCl products are equal on both sides of the membrane. Earlier, it was shown that alterations in the external KCl product produce KCl shifts and water movements in muscle that are characteristic for a Donnan system in osmotic equilibrium.[10] At high $[K]_o$ concentrations (5 mM and above), the Donnan law is obeyed well in muscle, whereas at lower values of $[K]_o$, the agreement is approximate only, unless account is taken of the permeability of the membrane to Na. A detailed treatment of osmotic theory is given in Chapter 3.

Active Transport

In contrast to the K distribution, which is low outside the membrane but high on the inside, the Na distribution is just the reverse. The value of $[\mathrm{Na}]_i$ is about 10 mM, whereas for $[\mathrm{Na}]_o$, it is 120 mM. This demonstrates the operation of a strong Na pump in muscle. Although detailed treatment of the sodium pump is deferred until Chapters 5 and 6, it is important at this point, for historical reasons, to highlight the classical observations on Na extrusion by skeletal muscle fibers.

Skeletal muscle fibers were allowed to gain Na and lose K by placing them for a length of time in K-free Ringer's solution. These muscles extruded the additional Na gained and reaccumulated the K that was lost when K ions at a suitable concentration were readmitted to the medium.[11-13] As these muscle fibers retained a resting membrane potential that was even more negative than normal, the Na extrusion was clearly against an electrochemical gradient.

The cardioactive steroids strophanthidin and ouabain have long been known to be inhibitors of Na transport. Skeletal muscle was the preparation used in elucidating the action of these inhibitors on Na transport as the application of these inhibitors prevented Na-enriched muscles from extruding Na in the presence of external K.[14,15]

Skeletal muscle cells formed a significant part of early studies on ion transport across cell membranes. Indeed, the term "sodium pump" appears to have been first used in the literature by Dean,[16] with reference to muscle cells. Dean also introduced the term "active transport," which, together with the term "pump," has found a permanent place in the literature. Dean's firm faith in thermodynamics led him to postulate the existence of a metabolically driven Na pump within the muscle fiber membrane itself, a pump that performs work at a constant rate to extrude Na ions as fast as they enter the fiber due to diffusion. He correctly saw that the K distribution in muscle fibers could be equally well explained by an inwardly directed K pump in addition to the Na pump or by solely an outwardly directed Na pump.[16] Despite the sparse kinetic and metabolic data available at the time, Dean's work contains all of the essential ingredients of modern concepts of the electrolyte distribution in muscle fibers. His work anticipates the "electrogenic Na pump" later discovered in muscle fibers[17-19] and sets the stage for practically all of the transport work on muscle fibers to follow.

Exchange Diffusion

There is a third type of ion movement in muscle that shares features with both electrodiffusion and active transport. Ussing[20] suggested that part of the Na efflux in frog muscle might be due to a mechanism in the fiber membrane that exchanges internal Na ions for external Na ions on a one-for-one basis. He termed the hypothesized process "exchange diffusion." Such a mechanism would require that Na efflux decline, upon removal of external Na ions from Ringer's solution, by an amount equal to the amount of Na flux participating in the 1:1 exchange. It was observed that about half the Na efflux in frog sartorius muscle disappeared upon replacing external Na with either Li or choline.[21] The exchange diffusion component of Na efflux in muscle fibers occurs in the presence of strophanthidin or ouabain and is therefore not a part of the ouabain-sensitive Na pump.[22,23] In a K-free medium, however, an Na:Na exchange occurs that is ouabain sensitive.[24,25] This type of exchange diffusion is abolished by addition of external K ions and seems to be a particular mode of Na pump behavior. The ouabain-insensitive type of exchange diffusion in muscle has not yet been elucidated further and, hence, must remain as a hypothetical separate possible mode of Na ion movement.

Categories of Ion Movement

Electrodiffusion, active transport, and exchange diffusion form three convenient categories in which to discuss ion movements in muscle and in other tissues as well. If an ion forms a neutral complex with a membrane-located carrier molecule that then obeys the laws of diffusion, a fourth category becomes necessary, namely, carrier-mediated or facilitated diffusion. The theory for carrier-mediated fluxes, pumped fluxes, and exchange diffusion is considered in later chapters. The laws of simple membrane diffusion and electrodiffusion belong in this introductory chapter, however, as they provide a basis for reference throughout the text.

BASIC THEORY FOR PASSIVE ELECTRODIFFUSION

By virtue of its internal structure, every biological cell membrane offers some resistance to the diffusive passage of a molecule or ion. The diffusion of an ion across a membrane is termed *electrodiffusion* as ion movement is sensitive to the electrical potential difference across the membrane as well as to the concentration gradient. The average velocity of the particles crossing the membrane is proportional to the force driving the particles across the membrane and inversely proportional to the resistance offered by the membrane. For diffusive movement across the membrane, the transmembrane flux is equal to the membrane concentration of particles at each point times the velocity at each point. The absolute mobility is the velocity per unit force; hence, flux = concentration × mobility × force. It is usual in flux theory to replace the forces by appropriate scalar quantities that are the negatives of the corresponding gradients in free energy. For the diffusion of a neutral particle, the free energy gradient governing the flux is the gradient in chemical potential. For electrodiffusion, the gradient of energy governing the flux is the gradient in electrochemical potential. The fundamental relation for flux is

$$j = -C_m U \nabla \psi \tag{2}$$

where j is flux in moles/cm^2·sec, C_m is membrane concentration, U is absolute mobility in cm/sec·dyne, and $\nabla \psi$ refers to the gradient in potential energy. For diffusion of a neutral molecule, ψ is the chemical potential, $\mu = \mu_o + RT \ln \gamma C$ where R is the gas constant per mole, T is the absolute temperature, C is concentration, and γ is the activity coefficient. If γ does not change across the membrane, $\nabla \mu = RT\, d\ln C/dx$ and equation (2) becomes, for diffusion in the X direction,

$$j = -RTU \frac{dC_m}{dx} \tag{3}$$

For steady-state diffusion, the concentration gradient must be linear and dC_m/dx becomes

$$\frac{(C_{m_i} - C_{m_o})}{\delta}$$

where i refers to the inner membrane surface, o to the outer membrane surface, and δ is the membrane thickness. From the theory of diffusion, the diffusion coefficient, D, is given by $D = RTU$. Making these substitutions, one obtains for the membrane diffusion of a neutral molecule

$$j = \frac{D}{\delta}(C_{m_o} - C_{m_i}) \tag{4}$$

As C_m, the membrane concentration, is not a directly measurable quantity, a partition coefficient has been defined as $\beta = C_{m_i}/C_i = C_{m_o}/C_o$ where C_i and C_o now refer to solution concentrations.[26] Defining the permeability coefficient, P, by the relation

$$P = \frac{\beta D}{\delta} \tag{5}$$

the equation for simple membrane diffusion becomes

$$j = P(C_o - C_i) \tag{6}$$

In the case of electrodiffusion of ions, ψ in equation (2) is electrochemical potential. The resulting equation for the flux of an ion across the membrane is

$$j = -C_m U \nabla \mu - zC_m UF \nabla E \tag{7}$$

where additional terms are: z, valence; F, the Faraday; and E, the electrical potential. This is the Nernst-Planck differential equation for flux. For ion movement in the X direction only, the equation is

$$j = -\frac{RTu}{|z|F}\frac{dC_m}{dx} - \frac{zC_m u}{|z|}\frac{dE}{dX} \tag{8}$$

where U is replaced by the electrical mobility given by $u = |z|FU$.

As applied to biological membranes, the potential gradient has been assumed to be constant. On the basis of this assumption, "constant field" solutions to the Nernst-Planck equation have been obtained.[26,27] The form to be employed in the text is

$$j = P\frac{zE_mF}{RT}\left(\frac{C_i e^{zE_mF/RT} - C_o}{e^{zE_mF/RT} - 1}\right) \tag{9}$$

where E_m is the membrane potential. This equation can be separated into two parts, *unidirectional efflux* and *influx*, if such fluxes are independent (see equations 54 and 55). Thus $j = j_o - j_i$.* A test of equation (9) for unidirectional K fluxes in frog sartorius muscle has been made.[8] The K fluxes vary with $[K]_o$ in such a way that it is possible to describe them using equation (9). When fluxes of other alkali metal cations in muscle are considered, however, it is necessary to amend the theory or select alternate theories. The subject is considered in Chapter 4.

If one forms the ratio of the unidirectional fluxes j_o/j_i on the basis of the independence assumption, the "flux ratio equation" is obtained:

$$\frac{j_o}{j_i} = \frac{C_i}{C_o} e^{zE_mF/RT} \tag{10}$$

In transport studies, the aim is to study individual components of flux unperturbed by the remaining fluxes. In practice, this is difficult to accomplish. In muscle, it is usually only possible to separate portions of flux modes from the remaining fluxes. For example, the K-sensitive and ouabain-sensitive portions of the Na pump can be readily separated experimentally from the remaining Na efflux, whereas the total active Na transport is difficult to separate in the absence of restrictive assumptions. It is also quite possible that the technique applied to study a particular kind of flux alters the remaining fluxes. The equations for electrodiffusion (9) and (10) are, therefore, helpful in separating diffusive or passive fluxes from active transport. There are ambiguities however. For an example, one can consider an ion that happens to be in electrochemical equilibrium. In this case both electrodiffusive and exchange diffusion components would obey equation (10) and theory would be of limited use in separating these flux components. Also, exchange diffusion and active transport have properties in common, making flux analyses difficult. Despite the difficulties, much progress has been made toward understanding the components of ion transport in muscle fibers. The theoretical background in this chapter should facilitate both further presentation of theory and discussion of experimental findings.

*For ion fluxes, a positive net flux is considered to be outward and, hence, an efflux. This convention is necessary for ions because they carry electrical current and because outward ionic current increases as the membrane potential is made more positive. For radioactive tracer kinetics, C_i increases when influx is positive. In this case the opposite convention is used so that $j = j_i - j_o$ (equation 20).

REFERENCES

1. W. O. Fenn and D. M. Cobb, *J. Gen Physiol.* **17,** 629 (1934).
2. R. H. Adrian, *J. Physiol.* **151,** 154 (1960).
3. R. A. Sjodin and L. A. Beaugé, *J. Gen. Physiol.* **61,** 222 (1973).
4. R. H. Adrian, *J. Physiol.* **133,** 631 (1956).
5. A. L. Hodgkin and P. Horowicz, *J. Physiol.* **148,** 127 (1959).
6. R. H. Adrian, *J. Physiol.* **156,** 623 (1961).
7. E. J. Harris, *J. Gen. Physiol.* **41,** 169 (1957).
8. R. A. Sjodin, *J. Gen. Physiol.* **42,** 983 (1959).
9. A. L. Hodgkin and P. Horowicz, *J. Physiol.* **145,** 405 (1959).
10. P. J. Boyle and E. J. Conway, *J. Physiol.* **100,** 1 (1941).
11. H. B. Steinbach, *J. Biol. Chem.* **133,** 695 (1940).
12. E. J. Conway and D. Hingerty, *Biochem J.* **42,** 372 (1948).
13. J. E. Desmedt, *J. Physiol.* **121,** 191 (1953).
14. P. A. Matchett and J. A. Johnson, *Fed. Proc.* **13,** 384 (1954).
15. J. A. Johnson, *Am. J. Physiol.* **187,** 328 (1956).
16. R. B. Dean, *Biol. Symp.* **3,** 331 (1941).
17. A. S. Frumento, *Science* (N.Y.) **147,** 1442 (1965).
18. L. J. Mullins and M. Z. Awad, *J. Gen. Physiol.* **48,** 761 (1965).
19. R. H. Adrian and C. L. Slayman, *J. Physiol.* **184,** 970 (1966).
20. H. H. Ussing, *Physiol. Rev.* **29,** 127 (1949).
21. R. D. Keynes and R. C. Swan, *J. Physiol.* **147**, 591 (1959).
22. R. D. Keynes, *J. Physiol.* **184,** 31P (1966).
23. L. A. Beaugé and R. A. Sjodin, *J. Gen. Physiol.* **52,** 408 (1968).
24. R. D. Keynes and R. A. Steinhardt, *J. Physiol.* **198,** 581 (1968).
25. R. A. Sjodin, *J. Gen. Physiol.* **57,** 164 (1971).
26. A. L. Hodgkin and B. Katz, *J. Physiol.* **108,** 37 (1949).
27. D. E. Goldman, *J. Gen. Physiol.* **27,** 37 (1943).

2

Kinetics and General Mechanisms of Ion Movements in Muscle

SINGLE-COMPARTMENT KINETICS

The net flux j of a substance into a single well-mixed cellular compartment separated from an outer solution by a thin membrane produces a time-dependent concentration change within the cellular compartment that is given by

$$\frac{dC_i}{dt} = \frac{A}{V} j \tag{11}$$

where A is the total surface area of the cellular compartment and V is the compartmental volume. For an electrically neutral solute, j can be substituted from equation (6). For the case of a constant external concentration C_o, provided physically by an infinite bath volume of external solution, the solution to the resulting differential equation is

$$C_i(t) = C_o(1 - e^{-kt}) \tag{12}$$

where the rate constant, k, is given by

$$k = \frac{A}{V} P \tag{13}$$

In cases where forces other than those of simple diffusion are involved, the terms multiplying C_o and C_i are not the same. Provided that the terms remain constant, however, the same general kinetic form is preserved. The differential equation to be solved is

$$\frac{dC_i}{dt} = k_i C_o - k_o C_i \tag{14}$$

For constant coefficients and constant C_o, the solution is

$$C_i(t) = \frac{k_i}{k_o} C_o(1 - e^{-k_o t}) \tag{15}$$

For electrodiffusion, the coefficients can easily be obtained using equation (9). For ion uptake in a single cellular compartment system,

$$C_i(t) = C_o e^{-zE_m F/RT}(1 - e^{-k_o t}) \tag{16}$$

where k_o is given by

$$k_o = P\left(\frac{A}{V}\right)\left(\frac{zE_m F}{RT}\right)\left(\frac{e^{zE_m F/RT}}{e^{zE_m F/RT} - 1}\right) \tag{17}$$

For electrodiffusion, it is clear that first-order kinetics can be expected only if the membrane potential remains constant during ion uptake or efflux. Also, equation (16) predicts that a steady state will be reached when the ion has reached electrochemical equilibrium (see equation 1).

In many applications, it is desirable to apply these kinetics to an ion already in a steady state, be it an electrodiffusional equilibrium or a pump-electrodiffusion (pump-leak) balance. In this case, use is made of a radioactive tracer that obeys the same kinetic law as the ion to be traced. The ratio of the number of radioactive particles to the total number of particles of that species is called the specific activity (S.A.). The S.A. on the outside of the cell is C_o^*/C_o, where the asterisk refers to the radioactively labeled particles. The equation for uptake of label is then

$$(\text{S.A.})_i = (\text{S.A.})_o (1 - e^{-kt}) \tag{18}$$

where k is the rate constant for efflux. It is convenient to refer these kinetics to the respective unidirectional influxes and effluxes. From equations (11) and (14), it is clear that

$$j = \frac{V}{A} k_i C_o - \frac{V}{A} k_o C_i \tag{19}$$

$$j = j_i - j_o \tag{20}$$

where j_i is unidirectional influx and j_o is unidirectional efflux. For constant C_o or $(\mathrm{S.A.})_o$, influx is constant and is obtained from the initial slope of the uptake. For the efflux of particles to an infinite bath of solution that effectively keeps the concentration of particles at a negligibly small concentration, equation (14) becomes

$$\frac{dC_i}{dt} = -k_o C_i \tag{21}$$

which integrates to

$$\ln \frac{C_i(t)}{C_{i_{t=0}}} = -k_o t \tag{22}$$

The same kinetic relation applies to the specific activity in the case of the movement of a radioactively labeled species. In this case, a plot of the logarithm of the fraction of radioactivity remaining in the cell against time should give a straight line with slope $-k_o$.

EXTRACELLULAR DIFFUSION

By convention used here, a single well-mixed cellular compartment in contact with an external solution of constant composition has been referred to as a single-compartment system. In many sources, such a situation is referred to as a two-compartment system, the extracellular phase forming a separate compartment. In most of the treatments made in this text, the composition of the extracellular fluid does not change during the experiment, which is why the choice has been made to generally exclude this compartment when discussing concentration changes within the muscle. As this procedure is an approximation that is often, but not always, valid, some discussion must be made concerning diffusion in the extracellular space. For a single muscle fiber, the extracellular space is confined to the thin unstirred layer immediately surrounding the fiber. Due to the thinness of this region, diffusional equilibrium is rapidly attained. When fibers are packed to form a whole muscle, however, diffusion in the extracellular region is slowed. The thinner the whole muscle, the less serious the problem. The sartorius muscle from the frog is a thin flat muscle less than a millimeter in thickness. Because of its relative thinness, the sartorius muscle has been widely used in transport studies.[1-4] To further avoid extracellular diffusional problems, small bundles of fibers from the frog semitendinosus muscle were employed by Horowicz and Gerber.[5]

Diffusional theory can be applied to correct for diffusional delays in whole muscle flux measurements.[1,6] In the model applied to uptake, the extracellular region gains a substance by inward diffusion from the extracellular solution in which the muscle is placed and loses the substance by membrane penetration of the substance into the fiber. Diffusion into the extracellular region is described by the diffusion equation in cylindrical coordinates:

$$\frac{\partial C_o}{\partial t} = D'\left(\frac{1}{r}\right)\frac{\partial}{\partial r}\left(r\frac{\partial C_o}{\partial r}\right) \tag{23}$$

where C_o is concentration in the extracellular space, D' is the diffusion coefficient in the extracellular space, and r is distance in the radial direction. The rate of loss of the substance due to membrane penetration is given by

$$\frac{\partial C_o}{\partial t} = -\left(\frac{1-\epsilon}{\epsilon}\right)\frac{\partial C_i}{\partial t} \tag{24}$$

where C_i is intrafibrillar concentration and ϵ is the volume fraction of extracellular space. Adding the two contributions to $\partial C_o/\partial t$ gives

$$(1-\epsilon)\frac{\partial C_i}{\partial t} + \epsilon\frac{\partial C_o}{\partial t} = \epsilon D'\left(\frac{1}{r}\right)\frac{\partial}{\partial r}\left(r\frac{\partial C_o}{\partial r}\right) \tag{25}$$

Using the previously presented relations between flux and $\partial C_i/\partial t$ and net flux and unidirectional fluxes, the solution obtained by Keynes[1] is

$$\frac{j'}{j} = \frac{\lambda}{b}\tanh\frac{b}{\lambda} \tag{26}$$

where j' is the measured tracer flux, j is the true flux corrected for diffusional delay, b is half the muscle thickness, and λ is defined by

$$\lambda^2 = \frac{\epsilon}{(1-\epsilon)} \left(\frac{D'}{j}\right) \left(\frac{r}{2}\right) [C]_o \tag{27}$$

where r is fiber radius and $[C]_o$ is concentration in the bathing medium. Applied to K^+ fluxes in whole frog sartorius muscle, j'/j has values around 0.9, indicating that fluxes of K^+ are reduced about 10 percent from the ideal single-fiber values by diffusion delay in the extracellular space.

THE EFFECT OF MULTIPLE CELLS ON KINETICS

A whole muscle contains several hundred individual muscle fibers with a range of fiber diameters. If all fibers had the same rate constant and if each fiber obeyed the first-order kinetics described, the whole muscle would follow kinetics given by the sum of all fibers' individual identical first-order kinetic terms. The whole muscle sum would itself be a first-order kinetic equation of the form given in equations (12) and (16). Even when given the same permeability coefficient for each individual fiber, however, the rate constants will differ if the fiber diameters differ due to the relation cited in equation (13). When there is large variation in the fiber diameters, the whole muscle does not invariably follow the same kinetic law that governs the individual fibers. Creese et al. have discussed this problem[7] and conclude that some differences in kinetic behavior can be explained on the basis of varying fiber diameters. However, it has been pointed out that K movement in the whole sartorius muscle would not be perturbed very much from individual fiber kinetics. Calculations show that a distribution of muscle fibers can be closely approximated kinetically if the entire distribution is thought of as having been replaced by an assembly of fibers all having the same diameter, a diameter just greater than the most frequent value in the distribution.[2]

This possible source of kinetic variation could, of course, be eliminated by working with single isolated muscle fibers.[8] The technical difficulties presented by the single striated muscle fiber preparation in tracer flux and transport studies are considerable, however. The literature survey to be made in this text shows that little would be known about transport in skeletal muscle if researchers had been forced to rely solely on single-fiber studies. As whole-muscle and small-bundle-of-fiber studies have contributed most of the information on transport in muscle, kinetic considerations are of some importance.

POTASSIUM FLUXES IN MUSCLE

The early studies of K movements in muscle using radioactive K ions as tracers are confusing as different authors have come to different conclusions. Muscle obtained from the toe of the frog (extensor longus dig. IV muscle) showed a linear ^{42}K output on a semilogarithmic plot versus time for a period of over two hours in agreement with equation (22). Influx of K was in approximate balance with efflux, and both fluxes increased in rough proportion to the value of $[K]_o$.[1] In rat diaphragm muscle, the loss of labeled K was linear on a semilogarithmic plot for about 1.5 hours. At longer times the rate of loss slowed somewhat.[7] Also, the value of the rate constant k_o (equation 15) determined from exchange during uptake showed good agreement with the rate constant determined from tracer washout, thus confirming the predictions of equation (15). Sartorius muscle from the frog, however, gave evidence of an initial fast fraction of ^{42}K uptake that could not be attributed to a dispersion of fiber sizes.[9] The effect was attributed to a region in the muscle fiber in which Na content is higher and K content lower than in the rest of the fiber. The fast fraction is interpreted by these authors as an exchange of external ^{42}K for K in the special region of the fiber.

Evidence for an inhomogeneity in the distribution of muscle K and Na was advanced by Harris and Steinbach,[10] who found that the specific activities of labeled K and Na ions lost from muscles to either K-free water or sugar solutions declined steadily with time during extraction. For a single fiber in which tracer ions are well mixed in the interior, tracer ions should be extracted at constant specific activity by solutions not containing these ions. The results indicate a greater degree of exchange of tracer K and Na in some parts of the whole muscle than in other parts. Unfortunately, such studies do not pinpoint the inhomogeneity, as some fibers could be more completely exchanged than others.

The muscle fiber model based on an inhomogeneous distribution of K and Na was pursued further by modifying the permeability kinetic laws summarized by equations (15) and (16) to include the effects of slow intracellular diffusion and a resistive coating surrounding each fiber.[2] The resulting equation for combined permeation and slowed intracellular diffusion can be summarized by stating the fact that, after a delay, ion uptake rises linearly with the square root of time. Another finding that is incompatible with the permeability model based on the assumption of complete intracellular mixing is the phenomenon of incomplete exchange of labeled isotopes with the intracellular species being traced.[10,11]

Kinetic anomalies could influence the interpretation of isotopic flux studies in the investigation of transport in muscle. In the case of ion efflux, one can sometimes ignore the early, non-first-order washout of an isotope and obtain the rate constant from the linear portion of a semilogarithmic plot. If uptake of an ion departs from first-order kinetics, however, it is difficult to relate membrane-limited influx to efflux. In studies of transport in muscle, one hopes to study events taking place in the fiber membrane. Therefore, kinetics of ion movements do not play a trivial role in transport studies in muscle unless they have been demonstrated to obey the first-order laws developed for membrane-limited movement.

The sartorius muscle from the frog has been a convenient and important source of information on ion transport. Before this preparation could be adopted for studies of ion transport across the fiber membrane, it was necessary to show that, under proper conditions, one could use the frog's sartorius muscle to study membrane-limited fluxes.[12] For frog sartorius muscle in or very near a steady state with respect to K and Na content, K fluxes obeyed equations (16) and (22) to within experimental error. The uptake of labeled K into frog sartorius muscle for a four hour period in Ringer's solution with $[\mathrm{K}]_o = 5$ mM is illustrated in Figure 2.1. The points are experimental values, and the smooth curve is a plot of the equation $C_i = C_i\infty$ $(1 - e^{-k_o t})$ with $C_i\infty = 80$ μmoles/g and $k_o = 0.20$ hr^{-1}. The value of $C_i\infty$ was obtained by flame photometry at the end of the experiment, and k_o is the rate constant determined from the washout of ^{42}K. The value of C_i indicates that the muscle was in, or very near to, a steady state. The curve strikes the uptake axis at a value of K uptake that can be accounted for by extracellular K with an extracellular space of 20 percent of the muscle volume. When uptake was followed for long times, tracer exchange proceeded to completion or very nearly so.

The washout of ^{42}K from a frog sartorius muscle is illustrated in Figure 2.2. An initial ^{42}K washout occurs that can nearly all be accounted for by the ^{42}K residing in the extracellular space. For this particular muscle, tracer exchange was only about 1 percent complete prior to washout. The kinetics and the rate constant evaluated did not differ appreciably from those obtained in cases where tracer equilibration was much more complete. In more complicated models for ion washout from muscle, different efflux kinetics are ex-

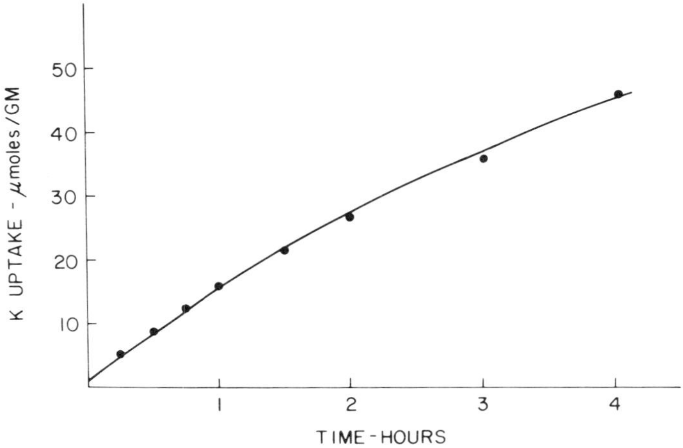

Figure 2.1. Plot of ^{42}K uptake in frog sartorius muscle versus time in contact with radioactive Ringer's solution containing $[K]_o$ = 5 m*M*. From data of R. A. Sjodin and E. G. Henderson, unpublished results.

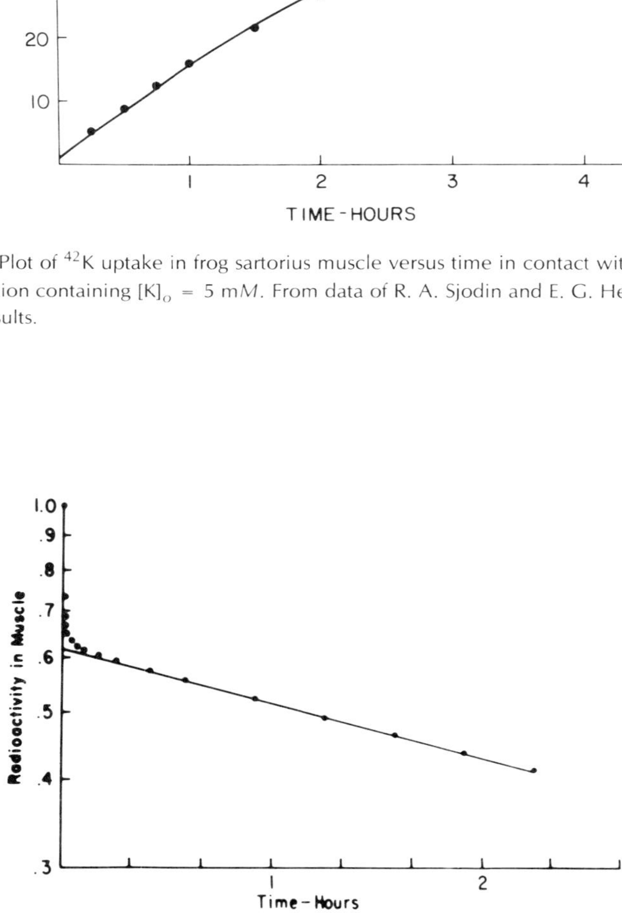

Figure 2.2. Radioactive K remaining in sartorius muscle from the frog during efflux is plotted against time semilogarithmically. The muscle had been previously loaded with ^{42}K ions for a period of 5 minutes, which gave 1.33 percent equilibration with tracer. $[K]_o$ = 2.5 m*M*. From R. A. Sjodin and E. G. Henderson, *J. Gen. Physiol.* **47,** 605 (1964).

pected when lightly equilibrated muscles are compared with more heavily equilibrated muscles. The decline in ^{42}K specific activity occurring when both ^{42}K and K are collected from labeled muscles in K-free Ringer's solution is plotted against time in Figure 2.3. The results are consistent with a fairly uniform distribution of ^{42}K within the sartorius muscle. The more heavily labeled early washout could indicate ^{42}K coming from a population of more heavily labeled fibers or from a more highly equilibrated region of each muscle fiber. Whatever the source of the nonuniformity in ^{42}K distribution in the muscle, it is not of sufficient magnitude to preclude treating the whole sartorius muscle as a first-order kinetic system with regard to the fiber K.

Non-first-order kinetics can be expected when the muscle is not in a steady state. These cases are considered in greater detail in Chapters 5 and 6. The main causes of non-steady-state kinetics are loss of fiber K, gain of fiber Na, and the reverse of these changes during net Na pumping. When the muscle Na becomes even slightly elevated, external K can exchange rapidly with the gained Na, giving rise to two fractions of K uptake. Earlier discrepancies in ion transport kinetics in muscle might be explainable on such a basis.[9]

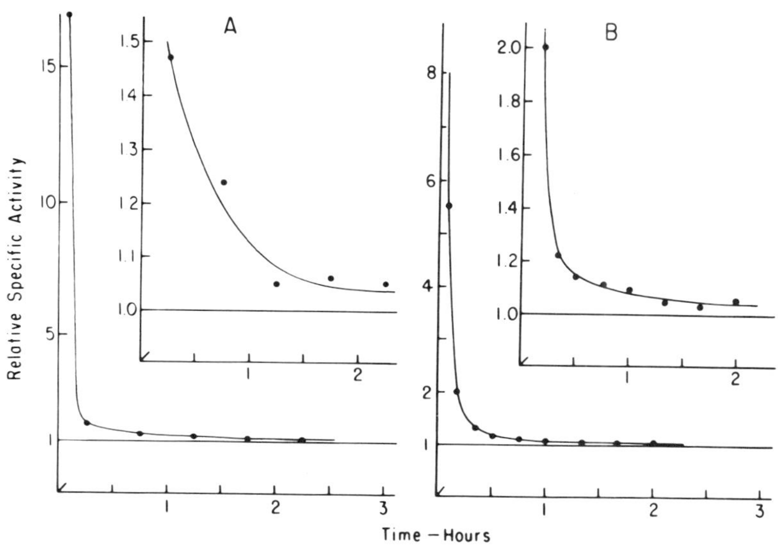

Figure 2.3. The relative specific activity of K ions lost from sartorius muscles of the frog to an originally K-free Ringer's solution is plotted against the time of collection of ^{42}K and total K. Graphs *A* and *B* are from two different muscles. The insets show expanded scales of specific activity. From R. A. Sjodin and E. G. Henderson, *J. Gen. Physiol.* **47,** 605 (1964).

Individual ion movements are considered in detail in chapters devoted to specific ions. Only the most general kinds of mechanisms for membrane penetration are considered here. A summary of K flux values for muscle membrane is given in Table 2.1. Despite some variation, there is general agreement in the results. For the same muscle type, increasing $[K]_o$ increases both K influx and efflux. Also, the K fluxes are nearly in balance except for the single-fiber case, which shows a considerable net efflux of K. For the single-fiber case, however, not enough fibers have been studied to give precise average values. It is more valid to compare values of K influx and efflux measured on the same fiber. For two such cases, the values are: K influx = 3.7, K efflux = 4.0; K influx = 6.4, K efflux = 6.5 (all values in pmole/ $cm^2 \cdot sec$).[8] The two resulting values for the flux ratio are 1.08 and 1.02, providing further evidence that the K fluxes in muscle are nearly in balance in Ringer's solution with nearly normal K concentrations.

It is of interest to compare the measured flux ratios for K with those calculated from equation (10). The calculated ratios for different K concentrations and membrane potentials are as follows:

$[K]_o$, mM	$[K]_i$, mM	E_m, mV	j_o/j_i (calculated)	j_o/j_i (measured)
2.5	140	−92	1.45	1.20
5.0	140	−80	1.17	1.02

Although the calculated values appear superficially to agree with the measured ratios, in fact, they are slightly, but significantly, higher. This

Table 2.1. Potassium Fluxes in Muscle Fibers

$[K]_o$, mM	j_i, pmole/$cm^2 \cdot$sec	j_o, pmole/$cm^2 \cdot$sec	j_o/j_i	Source
2.5	6.2	7.4	1.19	Sartorius, frog[12]
2.5	5.4	8.8	1.63	Single-fiber, frog[8]
2.5	4.1	4.9	1.20	Extensor longus
5	7.3	8.6	1.18	dig. IV, frog[1]
5	12.2	12.5	1.02	Sartorius, frog[12]
4.5	4.6	4.6	1.00	Digitorum longus, rat[13]

small difference is crucial because it indicates that muscles do not lose K to the extent predicted by electrochemistry alone. These results are consistent with the simultaneous operation of an inwardly directed K pump. This subject receives more detailed examination in Chapters 4–6. As it is difficult to rule out K:K exchange diffusion, which by definition has a flux ratio of 1.0, it is impossible to draw firm conclusions. One can only say that electrodiffusion and/or passive facilitated K diffusion probably accounts for most of the K fluxes in normal resting muscle.

SODIUM FLUXES IN MUSCLE

The loss of labeled Na from muscle occurs in at least two phases. The first is a rapid diffusion from the extracellular space, and the second is a slower exponential loss.[1,6,14] The slower loss can sometimes be fit by a single exponential, but it frequently shows a flattening of the semilogarithmic plot of the fraction of labeled Na in the muscle against time.[1] The presence of more than one exponential makes it difficult to speak of a definite transmembrane flux unless one of the exponentials can be shown to represent diffusion solely from the extracellular space. A technique that is often applied is to reckon the rate constant for cellular Na loss from the slowest and longest linear portion of a semilogarithmic plot. This technique is valid unless the Na in muscle fibers is *not* homogeneously distributed as other workers have suggested.[9-11] For fresh normal frog sartorius muscle, the rate constant for cellular Na loss is between 1.2 hr^{-1} and 1.5 hr^{-1}.[6,14,15] Taking an average value of 1.35 hr^{-1} and a cellular Na content of 6.6 μmoles/g,[16] Na efflux in fresh sartorius muscle from the frog is 5.8 pmoles/$cm^2 \cdot$sec using surface to volume data from Harris.[2] A value of 5.4 pmoles/$cm^2 \cdot$sec was deduced by Keynes using slightly different sources of data.[1] Due to the large fraction of Na residing extracellularly, it is difficult to make precise measurements of Na influx in muscle by the tracer method. Therefore, a different route must be taken to get an accurate estimate of the Na influx. The frog sartorius muscle bathed in normal Ringer's solution at 20°C is in nearly a steady state with respect to cellular Na for a period of at least 12 hours.[17] The net influx of Na is small and amounts to no more than around 0.1 pmole/$cm^2 \cdot$sec. There is, therefore, justification for taking Na influx for fresh frog sartorius muscle as also equal to about 5.8 pmoles/$cm^2 \cdot$sec.

The fractions of Na flux occurring in muscle are briefly mentioned here and examined in detail in Chapters 5 and 6. About one-half the Na flux in fresh sartorius muscle is due to exchange diffusion.[18] The remainder, 2.9 pmoles/cm^2·sec can be largely accounted for as an electrodiffusive inward leakage of Na ions.[16] Due to the large component of outward Na pumping in muscle fibers and the impossibility of completely unfueling the muscle Na pump, it is not feasible to accurately measure the electrodiffusive outward Na flux. As equation (10) appears to apply rather well to the electrodiffusive K fluxes in muscle, one might assume it applies to the Na fluxes as well and use it to calculate passive Na efflux from the Na concentrations, the membrane potential, and the passive Na influx. This technique yields a value of 0.007 pmole/cm^2·sec for the passive Na efflux. Thus almost all of the 2.9 pmoles/cm^2·sec of Na efflux must be due to the operation of the Na pump. This fraction of Na flux in muscle is augmented by increasing the value of $[K]_o$, by adding external K-like cations such as Cs, and by elevating the temperature.[1,19,20] The Na efflux is partially inhibited by the addition of cardioactive steroids externally.[19,20]

CHLORIDE FLUXES IN MUSCLE

The loss of labeled Cl from the frog's sartorius muscle to Cl Ringer's solution is linear on a semilogarithmic plot after a time in which loss from the extracellular space is largely complete.[21] There are difficulties in reckoning both $[Cl]_i$ and Cl efflux from such curves, however. The Cl efflux in muscles depends upon the nature of the external anion.[22] The loss of Cl is slowed in Ringer's solution in which Cl ions have been replaced by NO_3 ions. However, the initial slow rates of loss of ^{36}Cl measured after disappearance of the fast fractions are about the same for both Cl^- and NO_3^- Ringer's solution.[21] The intercepts at $t = 0$, however, which should represent the intracellular Cl contents, are considerably different. This finding suggests that there are kinetic difficulties associated with the measurement of Cl fluxes in muscle.

The Cl content of muscle, determined by extrapolating Cl loss into NO_3 Ringer's solution, lies between 3.1 and 3.8 mM. The time constant for ^{36}Cl loss to Cl Ringer's solution is about 11 minutes,[21] giving a rate constant of 0.09 min^{-1}. Using the same average surface to volume ratio employed for the other fluxes, the calculated value of Cl efflux in frog sartorius muscle is 8.4 pmoles/cm^2·sec. This value of Cl flux is severalfold too low to account for the expected contribution of Cl ions to the total membrane conductance. There is no certain explanation for the discrepancy between tracer fluxes and membrane conductance for Cl ions. A severalfold greater Cl efflux would be very difficult to separate kinetically from the washout of the extracellular space. One possible explanation for the discrepancy is that the transmembrane Cl efflux in fresh muscle with a normal Cl content simply cannot be measured in Cl Ringer's solution. An alternative explanation is that the tracer fluxes represent membrane fluxes but are lower than expected because of nonindependent Cl ion movement across the membrane.[21,23]

ION FLUXES IN SINGLE MUSCLE FIBERS

The main advantage in using single isolated muscle fibers in flux studies is that problems associated with the extracellular space and a dispersion of fiber sizes can be avoided. The main conclusion reached in an investigation of Na and K fluxes in isolated single semitendinosus muscle fibers from the frog is that the fluxes obey the single exponential kinetic relations for membrane-limited ion movements summarized in equations (15) and (22).[8] This fact is of itself important as it suggests that at least some of the kinetic anomalies previously discussed may be due to the extracellular space and to the dispersion of fiber size occurring in whole muscle. The mean values for K flux found in single fibers are given in Table 2.1, where they are comparable to fluxes for whole muscle. The closeness to values measured for whole muscle may indicate that large diffusional corrections are not required for the whole-muscle data. The value measured for Na flux in single fibers was 3.5 pmoles/cm^2·sec, compared to the value of 5.8 pmoles/cm^2·sec measured using whole sartorius muscle. Single-fiber fluxes are somewhat lower in magnitude, probably because whole-muscle fluxes involve a population of fibers, at least some of which have different characteristics. The single-fiber studies make use of the larger fibers, namely, 100 μ in diameter, which may have slightly different characteristics than the smaller fibers that contribute to whole-muscle fluxes. It should be noted that, for larger fibers with the same permeability characteristics, fluxes should be the same, whereas rate constants should be lower for the larger fibers in view of the relation given in equation (13). Rate constants should vary inversely with fiber radius. With the single fiber data, a test of this relation shows that within experimental errors, the relation holds.[8]

REFERENCES

1. R. D. Keynes, *Proc. Roy. Soc. B* **142,** 359 (1954).
2. E. J. Harris, *J. Gen. Physiol.* **41,** 169 (1957).
3. R. A. Sjodin, *J. Gen. Physiol.* **42,** 983 (1959).
4. R. H. Adrian, *J. Physiol.* **151,** 154 (1960).
5. P. Horowicz and C. J. Gerber, *J. Gen. Physiol.* **48,** 489 (1965).
6. E. J. Harris and G. P. Burn, *Trans. Faraday Soc.* **45,** 508 (1949).
7. R. Creese, M. W. Neil, and G. Stephenson, *Trans. Faraday Soc.* **52,** 1022 (1956).
8. A. L. Hodgkin and P. Horowicz, *J. Physiol.* **145,** 405 (1959).
9. M. J. Carey and E. J. Conway, *J. Physiol.* **125,** 232 (1954).
10. E. J. Harris and H. B. Steinbach, *J. Physiol.* **133,** 385 (1956).
11. E. J. Harris, *J. Physiol.* **120,** 246 (1953).
12. R. A. Sjodin and E. G. Henderson, *J. Gen. Physiol.* **47,** 605 (1964).
13. K. L. Zierler, E. Rogus, and C. F. Hazlewood, *J. Gen. Physiol.* **49,** 433 (1966).
14. H. Levi and H. H. Ussing, *Acta Physiol. Scand.* **16,** 232 (1948).
15. E. J. Harris, *Trans. Faraday Soc.* **46,** 872 (1950).
16. R. A. Sjodin and L. A. Beaugé, *J. Gen. Physiol.* **61,** 222 (1973).
17. L. J. Mullins and K. Noda, *J. Gen. Physiol.* **47,** 117 (1963).
18. R. D. Keynes and R. C. Swan, *J. Physiol.* **147,** 591 (1959).
19. C. Edwards and E. J. Harris, *J. Physiol.* **135,** 567 (1957).
20. L. A. Beaugé and R. A. Sjodin, *J. Physiol.* **194,** 105 (1968).
21. R. H. Adrian, *J. Physiol.* **156,** 623 (1961).
22. E. J. Harris, *J. Physiol.* **141,** 351 (1958).
23. A. L. Hodgkin and R. D. Keynes, *J. Physiol.* **128,** 61 (1955).

3

Electrolyte Distribution, Osmotic Equilibria, and Membrane Potentials

The high permeability of muscle membrane to both K and Cl permits KCl to move rapidly toward equilibrium. As these ions contribute significantly to the osmotic pressure inside the fibers, the KCl-movements are accompanied by water movement and volume changes. This differs sharply from the situation of nerve fibers and red blood cells, where low anion permeability in nerve fibers and low cation permeability in red blood cells prevent the rapid movement of salts.

The K, Na, and Cl contained in sartorius muscle fibers from the frog are presented in Table 3.1, where they are compared with Ringer's solution concentrations. The equilibrium potentials are calculated for each ion and compared with the resting membrane potential. The ion contents and membrane potential are from Adrian[1-3] and Sjodin and Beaugé.[4] The remaining constituents of Ringer's solution, which contribute only in a minor way to the osmotic pressure, are not included.

The data show that Cl is in electrochemical equilibrium in muscle to within the combined experimental errors in $[\mathrm{Cl}]_i$ and E_m, and that K ions are close to equilibrium. As muscle is in osmotic equilibrium, the sum of internal

Table 3.1. Ion Contents of Muscle

Ion	Inside concentration, mM	Ringer's solution, mM	E_{EQU}, mV	E_m, mV	$E_m - E_{EQU}$, mV
K	139[a]	2.5	−101	−92[b]	9
Na	10.6[c]	120	+61	−92	−153
Cl	3.5[d]	122.5	−90	−92	−2

[a] Data from R. H. Adrian, *J. Physiol.* **151,** 154 (1960).
[b] Data from R. H. Adrian, *J. Physiol.* **133,** 631 (1956).
[c] Data from R. A. Sjodin and L A. Beaugé, *J. Gen. Physiol.* **61,** 222 (1973).
[d] Data from R. H. Adrian, *J. Physiol.* **156,** 623 (1961).

K, Na, and Cl concentrations contribute slightly over half the total osmotic pressure. The major source of the required additional osmotic pressure is the internal nondiffusible anions. The electrolyte distribution in muscle thus approximates a Donnan equilibrium for KCl. When $[K]_o$ is elevated to 5 mM, the muscle K is very nearly in a steady state, E_K now being −84 mV and E_m −80 mV. Under these conditions, the approximation to a Donnan equilibrium permits the muscle fiber to be treated fairly adequately as a system in osmotic equilibrium and in electrochemical equilibrium with respect to KCl, as assumed by Boyle and Conway.[5]

OSMOTIC THEORY AND EQUILIBRIA

The osmotic relations for the muscle fiber may be determined by a method similar to that followed by Boyle and Conway. The condition for electrochemical equilibrium requires that

$$[K]_i[Cl]_i = [K]_o[Cl]_o \tag{28}$$

The condition for osmotic equilibrium requires that

$$[K]_i + [Cl]_i + \frac{P}{V} = C \tag{29}$$

where P is the number of moles of nondiffusible anion inside the fiber, V is the fiber volume, and C is the total osmotic concentration in the external medium. In the intial treatment, the value of $[Na]_i$ is ignored and the membrane is assumed to be impermeable to Na. This restriction will be removed subsequently. The condition for electrical neutrality inside the fiber demands that

$$[\mathrm{K}]_i = \frac{ZP}{V} + [\mathrm{Cl}]_i \tag{30}$$

where there are Z negative charges on every molecule of P. Solving these equations for $[\mathrm{K}]_i$ and $[\mathrm{Cl}]_i$ gives

$$[\mathrm{K}]_i = \frac{C}{2} + \frac{P}{V}\left(\frac{Z-1}{2}\right) \tag{31}$$

$$[\mathrm{Cl}]_i = \frac{C}{2} - \frac{P}{V}\left(\frac{Z+1}{2}\right) \tag{32}$$

Multiplying equations (31) and (32) and defining a parameter as given by

$$a = \frac{4[\mathrm{K}]_o[\mathrm{Cl}]_o}{C^2} \tag{33}$$

and solving for the volume gives

$$V = \frac{P[1 \pm \sqrt{1 + (Z^2 - 1)(1 - a)}\,]}{C(1 - a)} \tag{34}$$

For Ringer's solution of the kind indicated in Table 3.1 containing only Na, K, and Cl,

$$a = \frac{[\mathrm{K}]_o}{[\mathrm{Na}]_o + [\mathrm{K}]_o} \tag{35}$$

This equation is approximately valid for all Ringer's solutions containing only small quantities of additional constituents.

It is of interest to examine the solution to equation (34) when Z, the valence of P, is equal to 1.0. In this case

$$\frac{P}{V} = [\mathrm{Na}]_o \tag{36}$$

$$[\mathrm{K}]_i = \frac{C}{2} \tag{37}$$

$$[\mathrm{Cl}]_i = [\mathrm{K}]_o \tag{38}$$

Equation (36) states that under these conditions, only the nondiffusible entities determine the fiber volume. If Ringer's solution KCl composition is changed, keeping $[\mathrm{Na}]_o$ constant, the muscle fiber volume will remain constant. If Ringer's solution composition is changed, keeping the total concentration C constant, the volume will change, but the value of $[K]_i$ will remain constant. If the total concentration C and the product $[\mathrm{K}]_o[\mathrm{Cl}]_o$ are both held constant, there is no change in volume or internal concentrations even though $[\mathrm{K}]_o$ and $[\mathrm{Cl}]_o$ may vary reciprocally. The volume changes observed in muscle fibers when the composition of Ringer's solution is altered are consistent with these relations.[5] When the valence does not equal 1, the volume is obtained from equation (34) and $[\mathrm{K}]_i$ and $[\mathrm{Cl}]_i$ are calculated from equations (31) and (32).

INFLUENCE OF THE MEMBRANE POTENTIAL AND SODIUM ION PERMEABILITY

The main problem in the foregoing analysis is that the muscle fiber is not actually in a Donnan equilibrium with respect to KCl but only approximates such a situation at values of $[\mathrm{K}]_o > 5$ mM. If the Donnan condition held for muscle fibers, the resting membrane potential would be given by equation (1) for K ions as well as for Cl ions. In fact, however, the resting membrane potential has a more complicated dependence on $[\mathrm{K}]_o$, as shown in Figure 3.1, taken from Hodgkin and Horowicz.[6] These measurements are in Cl-free SO_4 Ringer's solution to avoid volume changes at high external K concentrations. Figure 3.1 further shows that equation (1) is strictly obeyed only at $[\mathrm{K}]_o = 10$ mM and above. The deviation observed below $[\mathrm{K}]_o = 10$ mM is that expected from a system that shows some leakiness to Na ions. If muscle fibers are not in a steady state with respect to Na and K but are with respect to Cl, the inward leak of Na will be just balanced by the outward leak of K. If the two ion leaks are passive fluxes obeying the electrodiffusion equation (9), the resting potential can be calculated by setting $j_{\mathrm{Na}} = -j_{\mathrm{K}}$. Solving the resulting equation for E_m gives

$$E_m = \frac{RT}{F} \ln \frac{[\mathrm{K}]_o + \alpha[\mathrm{Na}]_o}{[\mathrm{K}]_i + \alpha[\mathrm{Na}]_i} \tag{39}$$

where

$$\alpha = \frac{P_{\mathrm{Na}}}{P_{\mathrm{K}}} \tag{40}$$

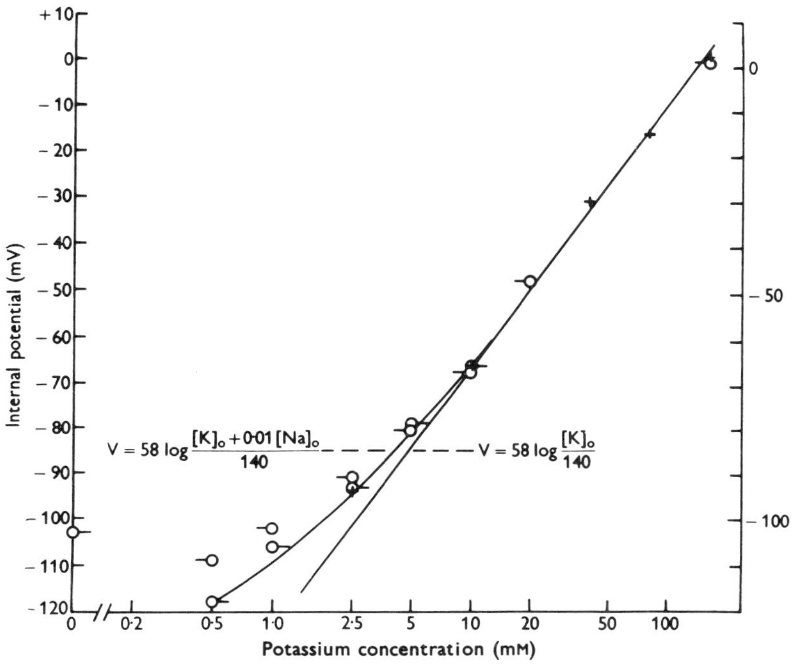

Figure 3.1. Relation between membrane potential of frog skeletal muscle fibers and log $[K]_o$ when using chloride-free sulfate Ringer's solution containing 8 m*M* $CaSO_4$. Crosses are potentials measured after equilibrating for 10–60 minutes; circles are potentials measured 20–60 seconds after a sudden change in concentration, –○ after increase in $[K]_o$, ○– after decreases in $[K]_o$. From A. L. Hodgkin and P. Horowicz, *J. Physiol.* **148,** 127 (1959).

A plot of equation (39) with $\alpha = 0.01$ gives an adequate fit to the data of Hodgkin and Horowicz.[6]

A finite permeability of the muscle fiber membrane to Na must be considered in modifying the osmotic theory of Boyle and Conway.[2,7] Equations (29) and (30) can be rewritten to include $[\mathrm{Na}]_i$:

$$[\mathrm{K}]_i + [\mathrm{Na}]_i + [\mathrm{Cl}]_i + \frac{P}{V} = C \tag{41}$$

$$[\mathrm{K}]_i + [\mathrm{Na}]_i = \frac{ZP}{V} + [\mathrm{Cl}]_i \tag{42}$$

As the membrane potential is given by both equations (39) and (1) for Cl ions,

$$([\mathrm{K}]_o + \alpha[\mathrm{Na}]_o)[\mathrm{Cl}]_o = ([\mathrm{K}]_i + \alpha[\mathrm{Na}]_i)[\mathrm{Cl}]_i \quad [43]$$

Equations (41) and (42) may be solved to give $[\mathrm{K}]_i$:

$$[\mathrm{K}]_i = \frac{(1 - Z)[\mathrm{Cl}]_i - (1 + Z)[\mathrm{Na}]_i + ZC}{1 + Z} \quad (44)$$

Substitution from equation (43) results in a quadratic equation for $[\mathrm{Cl}]_i$:

$$[\mathrm{Cl}]_i = -W \pm \sqrt{W^2 + X([\mathrm{K}]_o + \alpha[\mathrm{Na}]_o)} \quad (45)$$

where

$$W = \frac{(\alpha - 1)(1 + Z)[\mathrm{Na}]_i + ZC}{2(1 - Z)} \quad (46)$$

and

$$X = \frac{1 + Z}{1 - Z}[\mathrm{Cl}]_o \quad (47)$$

Internal K and Cl concentrations, as well as the relative fiber volume, can be calculated from a knowledge of external concentrations, $[\mathrm{Na}]_i$, Z, and α.

The measured values of $[\mathrm{K}]_i$ and $[\mathrm{Cl}]_i$ in frog muscle equilibrated in different Ringer's solutions agree well with the predictions of equations (44) and (45) with $Z = 1.63$.[2] It should be noted that the value of Z in Adrian's treatment[2] is -1.63, whereas Z has a positive value in the equations developed in the present text. This is because Z is here defined simply as the number of negative charges on the nondiffusible anion, P.

Measured volume relations in muscle agree with the theory outlined in this chapter. When $[\mathrm{K}]_o$ is increased at constant $[\mathrm{Cl}]_o$ by substitution for $[\mathrm{Na}]_o$ to hold C constant, $[\mathrm{K}]_i$ does not change but the muscle fibers swell due to KCl entry, even though the total osmotic pressure of the external medium remains constant. When $[\mathrm{K}]_o$ and $[\mathrm{Cl}]_o$ are increased by adding solid KCl to Ringer's solution and holding $[\mathrm{Na}]_o$ constant, $[\mathrm{K}]_i$ and $[\mathrm{Cl}]_i$ increase but the fiber volume remains constant[1,5] (refer also to equations 36 to 38). On the other hand, when $[\mathrm{K}]_o$ and $[\mathrm{Cl}]_o$ are varied at constant $[\mathrm{K}]_o[\mathrm{Cl}]_o$ product, the Boyle and Conway theory predicts no KCl entry, no volume change, and no change in either $[\mathrm{K}]_i$ or $[\mathrm{Cl}]_i$. It is expected that under these conditions, the membrane potential will rapidly reach a new constant value when $[\mathrm{K}]_o$ and

$[Cl]_o$ are changed. These expectations are borne out in experiments on frog single muscle fibers[6] and on crayfish muscle fibers.[8] Finally, in cases where $[Cl]_o$ is replaced with a nonpermeating anion such as SO_4, no volume changes occur when $[K]_o$ is varied and the membrane potential rapidly responds to changes in $[K]_o$ by reaching a new steady value.[1,6]

STEADY-STATE ELECTROLYTE DISTRIBUTION DUE TO PUMP LEAK BALANCE

The early approach to electrolyte balance in muscle assumed that the K and Cl of the fiber were in electrochemical equilibrium. Were this the case for all conditions, clearly no additional forces would be necessary to explain the distribution of K and Cl. However, as stated in Chapter 1, the low value of $[Na]_i$ can be attributed to the presence in the membrane of a metabolically driven Na pump. Such a pump could maintain the K distribution in two ways, either by assuring that the resting membrane potential, E_m, is equal to the K equilibrium potential, E_K, or by coupling the extrusion of Na to the inward pumping of some K ions. An indication of which mode of operation actually exists can be found in Table 3.1, where E_m is shown to be 9 mV less negative than E_K. In the treatment of Hodgkin and Horowicz,[6] the departure of E_m from E_{K} is attributed to an inward resting leak of Na ions.

Equation (39) fits well with data obtained from excised muscle, which is not necessarily in a steady state. In the physiologically intact animal, however, a steady state must prevail, and any electrodiffusive resting leaks of K and Na must be balanced by equal and opposite pumped ion movements. The only alternative is to suppose that in the intact animal, the values of E_m and E_{K} correspond, so that the electrodiffusive leakage of K ions disappears.

Another factor that should be taken into account is that, due to the thermodynamic activity coefficients of ions within the fiber, all equations for the membrane potential should, for the sake of accuracy, include the product of activity coefficients and ionic concentrations. The justification for omitting activity coefficients in almost all biological treatments is that they are unknown for the cell interior in most instances, and there is no reason to assume that they differ greatly from those in the external medium. In most cases, the omission of activity coefficients probably does not offer a source of error sufficient to alter qualitative conclusions. In cases where E_m differs from E_{K} by only a few millivolts, as in muscle fibers, however, the question of activity coefficient becomes more serious. For example, the difference between E_m and E_K in frog muscle fibers is reduced by taking the activity coefficients for K ions into account.[9]

An electrically neutral Na-K pump that transports K ions inwardly at the same rate that Na ions are transported outwardly has no direct effect on the membrane potential. In this case, the concepts underlying equation (39) are sufficient to describe the membrane potential. In the event that the pump is not electrically neutral, the pumped fluxes must be taken into account.[9] To achieve a steady state, the electrodiffusive leakage fluxes of both K and Na (equation 9) must be balanced by equal and opposite respective pumped fluxes. Denoting the pumped fluxes of Na and K, respectively, by m_{Na} and m_K,

$$j_{Na} = -m_{Na} \tag{48}$$

$$j_K = -m_K \tag{49}$$

Defining the coupling ratio r as the number of Na ions pumped for every K ion pumped,

$$r = -\frac{m_{Na}}{m_K} \tag{50}$$

Dividing equation (48) by equation (49) and substituting from equation (50) gives

$$j_{Na} = -rj_K \tag{51}$$

Substituting values of j_{Na} and j_K using equation (9), and solving for the membrane potential results in the following relation as obtained by Mullins and Noda[9]:

$$E_m = \frac{RT}{F} \ln \frac{rP_K[\mathrm{K}]_o + P_{Na}[\mathrm{Na}]_o}{rP_K[\mathrm{K}]_i + P_{Na}[\mathrm{Na}]_i} \tag{52}$$

Recalling that $P_{Na}/P_K = \alpha$,

$$E_m = \frac{RT}{F} \ln \frac{r[\mathrm{K}]_o + \alpha[\mathrm{Na}]_o}{r[\mathrm{K}]_i + \alpha[\mathrm{Na}]_i} \tag{53}$$

It is easily verified that equation (39) is obtained as a special case only if $r = 1$ and the pump is electrically neutral. Another limiting case occurs when no K ions are pumped at all, in which case $r \rightarrow \infty$ and $E_m = E_K$. The value of r for muscle fibers can be determined from measurements of pumped Na and

K fluxes (to be discussed in Chapters 5 and 6) or it can be evaluated from membrane potential measurements if α is known accurately from other independent measurements. From previous measurements of the membrane potential, one only knows that $\alpha/r = 0.01$.[6,9] If electrodiffusive leakage rates for K and Na are known, P_{Na} and P_K can be determined from equation (9) for Na and K, respectively, and the value of α determined. Using Na leakage rates from Sjodin and Beaugé,[4] and K leakage rates from Sjodin and Henderson[10] and Sjodin and Beaugé,[4] the value determined for α is 0.02, giving a value of $r = 2.0$. The value of r obtained from measurements of pumped fluxes is discussed in Chapter 6.

REFERENCES

1. R. H. Adrian, *J. Physiol.* **133,** 631 (1956).
2. R. H. Adrian, *J. Physiol.* **151,** 154 (1960).
3. R. H. Adrian, *J. Physiol.* **156,** 623 (1961).
4. R. A. Sjodin and L. A. Beaugé, *J. Gen. Physiol.* **61,** 222 (1973).
5. P. J. Boyle and E. J. Conway, *J. Physiol.* **100,** 1 (1941).
6. A. L. Hodgkin and P. Horowicz, *J. Physiol.* **148,** 127 (1959).
7. A. L. Hodgkin, *Proc. Roy. Soc. B.* **148,** 1 (1958).
8. J. Zachar, D. Zacharová, and M. Henček, *Physiol. Bohemoslav.* **13,** 177 (1964).
9. L. J. Mullins and K. Noda, *J. Gen. Physiol.* **47,** 117 (1963).
10. R. A. Sjodin and E. G. Henderson, *J. Gen. Physiol.* **47,** 605 (1964).

4

Mechanisms of Potassium Ion Movement in Muscle

Potassium ion permeability is related both to K flux and the membrane potential. Individual ion fluxes are related to specific ionic membrane conductances. The sum of these conductances, for all ions that can cross the membrane, constitutes the total membrane conductance. The purpose of this chapter is to describe the relationships that have been observed for net K fluxes, unidirectional K fluxes measured with ^{42}K, P_K, and the membrane conductance to K ions.

MEMBRANE PERMEABILITY TO POTASSIUM IONS

Given the net K flux, the membrane potential, and the K concentrations, equation (9) can be used to calculate P_K. The value of P_K varies with the driving force for K movement.[1-3] When E_m is close to E_K, P_K is around 1 $\times$

10^{-6} cm/sec. When the driving force is inward, and E_m is far from E_K, P_K rises. Conversely, P_K falls with an increasing outward driving force. This behavior is in marked contrast to that of P_{Cl}, which remains at rather constant values independent of the driving force producing the movement.

To determine values of P_K from tracer unidirectional flux data, equation (9) can be used only if the concentration on one side of the membrane is made zero or if the unidirectional fluxes occur independently of one another. In these cases, equation (9) can be separated into two parts as follows:

$$j_o = P\frac{zE_mF}{RT}\left(\frac{C_i e^{zE_mF/RT}}{e^{zE_mF/RT} - 1}\right) \tag{54}$$

and

$$j_i = P\frac{zE_mF}{RT}\left(\frac{C_o}{e^{zE_mF/RT} - 1}\right) \tag{55}$$

where j_o is efflux and j_i is influx. Using unidirectional flux data for K, P_K varies little when E_m is close to E_K and is not a function of either E_m or $[K]_o$ over a wide range in these quantities.[4,5] The value of P_K under these conditions is about 0.7×10^{-6} cm/sec for both K influx and efflux. This differs little from values obtained from transport numbers determined from E_m changes or from net movements when E_m is close to E_K. Therefore, there is negligible discrepancy between tracer flux data and net flux data in this regard. Further, there does not seem to be any great deviation from the principle of independent ion movement under these conditions.

The contribution of a species of ion to the membrane conductance can be calculated from flux data. If equation (9) is multiplied by the valence times the Faraday to obtain ionic current density, and if the result is differentiated with respect to the membrance potential, an equation for the ionic conductance is obtained.[4,6] For the case where P is a function of E_m, the equation for the ionic conductance for monovalent ions is

$$G = \frac{F^2}{RT}\left\{j_i - (j_i - j_o)\left[\frac{RT}{E_mF} + \frac{1}{1 - e^{E_mF/RT}} + \frac{RT}{F}\frac{\partial \ln P}{\partial E_m}\right]\right\} \tag{56}$$

The value of G can be calculated from the membrane potential, the unidirectional fluxes, and the variation of P with E_m. For the special case of flux balance at equilibrium ($j_i = j_o$), the equation for the conductance becomes

$$G = \frac{F^2}{RT} j \tag{57}$$

where j is either influx or efflux. If the unidirectional fluxes of an ion are passive, independent, and electrodiffusive in nature, equations (56) or (57) should predict the contribution of the fluxes of that ion to the membrane conductance.

For the frog's sartorius muscle,[7] the membrane resistance is 4000 $\Omega \cdot \text{cm}^2$, which gives a total membrane conductance of 250 μmhos/cm^2. The total conductance consists mainly of contributions from Cl and K ions, Cl ions contributing about two-thirds of the conductance.[3] In normal Ringer's solution with $[\text{K}]_o$ = 2.5 mM, one would, therefore, expect the membrane conductance to K ions at rest to be around 80 μmhos/cm^2. However, using unidirectional K flux data from Table 2.1, equation (56) predicts a membrane conductance to K ions of no more than 30 μmhos/cm^2 with $[\text{K}]_o$ = 2.5 mM, assuming no steady-state change in P_K with E_m. The discrepancy is considerable and cannot be attributed to experimental errors. In another conductance study, Sperelakis et al. used Ba ions to block the K conductance and found G_Cl = 196 μmhos/cm^2 and G_K = 67 μmhos/cm^2 for frog sartorius muscle in normal Cl Ringer's solution.[8] Clearly, calculations of G_K from flux data are too low to agree with any of the electrical data.

When an ionic flux is in or near a steady state, equation (57) may be used to calculate the expected conductance. Muscles in 5 mM K Ringer's solution approximate this condition. Using K flux data from Sjodin and Henderson[5] for $[\text{K}]_o$ = 5 mM, equation (57) predicts a K conductance of 48 μmhos/cm^2. There do not appear to exist electrical measurements of G_K when $[\text{K}]_o$ = 5 mM. However, if P_K is constant and if $[\text{K}]_i$ does not change, it can be shown that the steady-state conductance is approximately proportional to $[\text{K}]_o$. Though P_K changes with driving force,[3] it does not change with $[\text{K}]_o$, if K ions are near a steady state.[4] Thus the K conductance should approximately double when $[\text{K}]_o$ is changed from 2.5 to 5 mM. Taking G_K in Ringer's solution with $[\text{K}]_o$ = 2.5 mM as the average between 80 and 67 μmhos/cm^2, and multiplying by 2 to get G_K when $[\text{K}]_o$ = 5 mM gives a value of 147 μmhos/cm^2. The value of G_K from steady-state flux calculation (48 μmhos/cm^2) is three times too low to show agreement between passive-independent flux data and electrical measurements of membrane conductance resulting from K ions. These calculations are summarized in Table 4.1. The general conclusion from the data presentation in Table 4.1 is that the unidirectional K fluxes in muscle are not independent.

Table 4.1. Comparison of Membrane K Conductances

	K conductance, μmho/cm^2		
$[K]_o$, mM	From measurement[a]	Calculated from flux[b]	Ratio, measured/calculated
2.5	74	27	2.7
5.0	147	48	3.1

[a]Conductance values are averaged from the data of A. L. Hodgkin and P. Horowicz, *J. Physiol.* **148,** 127 (1959) and N. Sperelakis, M. F. Schneider, and E. J. Harris, *J. Gen. Physiol.* **50,** 1565 (1967). See text for details.
[b]Flux values used are from R. A. Sjodin and E. G. Henderson, *J. Gen. Physiol.* **47,** 605 (1964). Conductance values are calculated from equation (56) for $[K]_o = 2.5$ mM and from equation (57) for $[K]_o = 5.0$ mM.

NONINDEPENDENT POTASSIUM FLUXES IN MUSCLE AND THE POTASSIUM FLUX RATIO

When unidirectional influxes and effluxes do not occur independently, it is not permissible to separate equation (9) into two unidirectional parts. Hence equation (10) cannot be obtained under these conditions. A physical model that will produce nonindependent unidirectional fluxes consists of diffusion in a narrow pore that does not permit passing of particles.[9] In this case, equation (10) does not apply to the flux ratio, but a similar equation is obtained with the right-hand side raised to the $n + 1$ power where n is the number of single-file particle positions in the pore[9,10]:

$$\frac{j_o}{j_i} = \left(\frac{C_i}{C_o} e^{zE_mF/RT}\right)^{n+1} \tag{58}$$

The equation for ionic membrane conductance in the steady state now becomes

$$G = (n + 1)\frac{F^2}{RT} j \tag{59}$$

If this model is applied to the data summarized in Table 4.1, it is clear that $n + 1$ must be around 3 and n about 2. As there is no proof for this particular mechanism, one can speak of an "equivalent long pore" and simply refer to n as the interaction factor. An "interaction factor" of 2 brings K flux data, as measured with isotope, into line with conductance measurements.

When the K fluxes are in balance, the flux ratio is given by either equation (10) or equation (58) as it is equal to unity. A good test of flux ratio relations to determine the degree of independence of fluxes cannot, therefore, be made near the steady state. When $[K]_o$ is reduced to very low values, K influx declines much more than does K efflux[10] and the ratio j_o/j_i reaches values around 9. Under these conditions, the flux ratio departs sufficiently from a value of unity to enable a good test of equations (10) and (58) to be made. Results agreed reasonably well with predictions of equation (10) but not with predictions of equation (58) with $n = 1$. If only integral or zero values of n are permitted, the K fluxes obey the independence principle. This test for independent K ion movement was made on the assumption that all of the K fluxes measured are electrodiffusive and that no K exchange diffusion or active K transport is present. Later work showed that approximately 5 percent of the K influx into muscles with normal values of [Na] is due to active K transport.[11] This correction applied to K influx fails to alter the conclusions.

The possible exchange diffusion of K is difficult to evaluate. Approximately one-half the K efflux disappears when K ions are removed from the external medium.[5] This K efflux reduction is accompanied by an approximately 20 mV hyperpolarization of the membrane, which just about exactly accounts for the observed flux change on the basis of equation (54), if P_K remains constant. The value of P_K determined from the measurement of K efflux under these conditions is 0.6×10^{-6} cm/sec, and this does not differ significantly from the previous value of P_K given for the steady state. There is no reason to suppose, therefore, that K exchange diffusion takes place to any significant extent in normal muscle fibers. However, the presence of even a small amount of K exchange diffusion could affect the measurements considerably at low values of $[K]_o$ where the magnitude of K influx is very low in absolute value. Unfortunately, there does not seem to be any way to rule out this factor, as skeletal muscle fibers cannot be perfused or dialyzed with a K-free medium. One can only set some limits on the extent to which K exchange diffusion would alter conclusions. For example, if half of the K fluxes occurring normally were due to exchange diffusion and if half of the K influx remaining at low $[K]_o = 0.2$ mM is due to exchange diffusion, n would be 0.5 in the above analysis. To give a value of $n = 1$ would require that over 80 percent of the K influx, when $[K]_o = 0.2$ mM, be an exchange diffusion.

To be consistent with a nonindependent flux model, a very special kind of

K exchange diffusion would have to exist; the process would have to have a very high affinity for external K ions and the maximum value would have to be a small fraction of the total K flux that occurs when $[K]_o = 2.5$ mM or greater. If this were not true, the disagreement between conductances to K ions calculated from flux and those determined from electrical measurements would become worse than the factor of 3 deviation evident in Table 4.1. As such disagreement can always be handled by assigning higher values to n; however, one must admit that a high degree of nonindependent electrodiffusive K ion movement and a large amount of K exchange diffusion could coexist to give a pseudo agreement with the flux theory for independent ion movement at low values of $[K]_o$. Given no experimental evidence for the existence of K exchange diffusion in muscle, however, these considerations must be regarded only as possibilities.

It has been stated that P_K falls to low values when the outward driving force for K movement becomes elevated. This fact can be exploited to produce a muscle fiber in which the membrane potential is controlled solely by the Cl concentration ratio and is uninfluenced by the K concentration ratio. When fibers are soaked in solutions with elevated KCl concentrations (95 mM), the membranes depolarize to a potential around -20 mV; as a result, the fibers achieve an elevated internal Cl concentration consistent with this potential. When such fibers are replaced in Ringer's solution with normally low values of $[K]_o$, they fail to repolarize.[3] The reason for the failure to repolarize is that when $[K]_o$ is lowered to normal values, E_K is now some 80 mV more negative than E_m, thus providing a large outwardly directed driving force for K ion movement. The value of P_K, already lower than P_{Cl} even under normal conditions, now falls to such low values that the membrane potential no longer responds to K concentration changes and is effectively clamped by the Cl concentration ratio. These chloride-clamped fibers are extremely useful in studying K ion fluxes. This is because fluxes can be measured as a function of $[K]_o$ at constant values of E_m, and they can also be studied at different values of E_m at constant values of [K] and E_K.

In a further test of the independence principle, Horowicz et al.[12] examined the dependence of the K flux ratio in frog muscle on the magnitude of $E_m - E_K$, when E_K is kept constant and E_m varies. It was found under these conditions that the measured K flux ratio obeyed equation (58) with $n + 1 = 2$. It should be pointed out that the n of Horowicz et al. is equivalent to the $n + 1$ of equation (58). Thus in their treatment, they found $n = 2$. The results were interpreted in terms of a nonindependent model for the K fluxes in muscle that involves second-order kinetics. These conclusions are quite different from those reached when $[K]_o$ is much lower than normal and E_m has values around -110 mV. In this case, the K fluxes obey equation (10), which was derived on the basis of independent ion movements.[10] In the experiments of Horowicz et al., the membrane potential was varied between values of -20 and $+20$ mV, and P_K was very low.[12] It is quite possible that the mechanism for the unidirectional K fluxes in muscle is altered considerably by the membrane changes that lead to the large diminution in P_K that occurs at large outward driving forces with more positive values of the internal potential.

In summary, the observed K flux ratio obeys the standard electrochemical equation for independent fluxes when the K fluxes are in or near the normal physiological steady state and when the K fluxes are considerably out of balance and E_m is around -110 mV. In highly depolarized fibers, however, the K fluxes become markedly nonindependent and follow higher than first-order kinetics.

The fact that simple electrochemical laws are sometimes obeyed by the K fluxes in muscle should not lead one to oversimplify the mechanistic nature of these fluxes. In all cases observed so far, there is marked disagreement between measured membrane conductance for K ions and K conductance calculated from the unidirectional fluxes.

KINETIC PROPERTIES OF UNIDIRECTIONAL POTASSIUM FLUXES

Unidirectional K fluxes in muscle fibers have been measured as a function of $[K]_o$ both at varying and at constant membrane potentials.[4,12,13] The results of these studies are summarized in Figure 4.1. The solid dots refer to K influx occurring at the particular value of E_m that is appropriate for each value of $[K]_o$.[4] As these muscle fibers are in approximately steady-state conditions for fiber K, the plotted points refer very nearly to K efflux as well. High values of $[K]_o$ were present in sulfate Ringer's solution to prevent swelling due to KCl entry and also to prevent a net K influx from occurring. The K flux values are in fair agreement with equations (54) and (55), with a constant value of P_K around 10^{-6} cm/sec. The open symbols are plotted from Horowicz et al.[12] Since this data was normalized to 1.0 in the original work, the right-hand scale applies. The value 1.0 was chosen to correspond with the maximum flux observed in the data illustrated by the solid points. The absolute values of these fluxes do not show equality at the maximum value. The purpose of the plots is to compare the kinetic shape of the curves obtained at

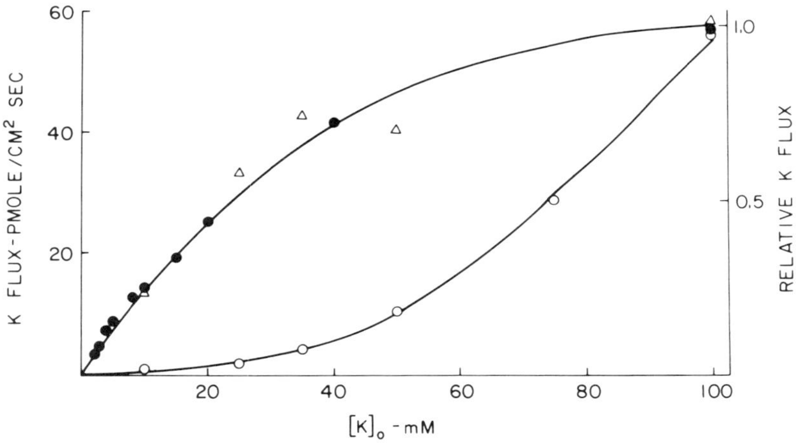

Figure 4.1. Potassium fluxes in frog skeletal muscle as a function of $[K]_o$. Solid dots refer to K influx in frog sartorius muscle at the variable E_m appropriate for each value of $[K]_o$; data from R. A. Sjodin, *J. Gen. Physiol.* **42,** 983 (1959). The open triangles refer to K efflux in frog semitendinosus muscle with E_m held constant at −18 mV and open circles refer to K influx under the same conditions; data from P. Horowicz et al., *J. Gen. Physiol.* **51,** 193s (1968); right-hand scale applies to all open symbols. All K fluxes measured using ^{42}K.

varying membrane potential (solid points) and at a constant membrane potential (open symbols) of −18 mV. The open triangles refer to K efflux and the open circles to K influx. The normalized K efflux at constant potential gives a good kinetic fit to the curve obtained at varying potential. The curve for K influx at a constant membrane potential of around −20 mV shows a marked departure from the upper curve and is proportional to the square of $[K]_o$. It is evident that it is the abnormal behavior of K influx at more positive internal potentials that causes the K flux ratio to deviate from the ideal relationship in the experiments of Horowicz et al.[12]

Some important clues concerning the nature of the K unidirectional fluxes in muscle fibers may be found in Figure 4.1. If the K fluxes follow the constant-field equation for flux (equations 9, 54, and 55), the downward bend of the upper curve for the solid points would be due to the depolarizing action of $[K]_o$ on the fiber membrane. At more positive interior potentials as $[K]_o$ rises, K influx would be expected to fall below a linear dependence on $[K]_o$ and K efflux at constant $[K]_i$ would be expected to rise due to the reduced holding power of the negative potential inside the fibers. There is nothing in the data of Sjodin that would indicate that this is not the correct explanation for the curve.[4] However, the fact that the efflux data of Horowicz et al.,[12] obtained at a constant membrane potential of -18 mV, can be made by normalization to fit the same curve casts considerable doubt on the foregoing application of the constant-field equation.

If the membrane depolarization occasioned by elevation of $[\mathrm{K}]_o$ is not the reason for the bend in the upper curve for the steady-state data (solid points), then $[\mathrm{K}]_o$ changes must of themselves produce the bend. A curve of the observed shape would be expected if the primary event in initiating a K flux is the reaction of a single external K ion with an externally directed site located in the fiber membrane.[13] If flux is proportional to the fraction of the membrane sites that have become occupied by K ions, one can write the equation for K influx as

$$j_i = \bar{j}\left(\frac{[\mathrm{K}]_o}{[\mathrm{K}]_o + \mathrm{K}_m}\right) \tag{60}$$

where $\bar{j}$ is the maximum flux that occurs when all the membrane sites are occupied by K ions and K_m is the reciprocal of the affinity of the sites for K. A satisfactory fit to the upper curve in Figure 4.1 is obtained with $\bar{j}_i = 86$ pmoles/cm$^2\cdot$sec and $\mathrm{K}_m = 43$ mM.[13] It should be noted that the shape of the curve described by equation (60) is determined only by the value of K_m. The maximal flux, $\bar{j}$, acts only as a scaling factor. For this reason, it is especially interesting that the normalized K efflux data of Horowicz et al.[12] fit the same curve with a half-maximal flux $[\mathrm{K}]_o$ of around 40 mM. In treating their own data, these authors required a much more elaborate model involving a mobile carrier within the membrane that associates with K ions in a second-order reaction. The membrane potential itself is visualized as arising from two surface potentials and one interior potential for the membrane proper. In addition, the carrier is activated by a nonmobile control site that associates with a single external K ion according to an equation similar in form to equation (60). When the control site is not associated with a K ion, the carrier is inactive even when it is in the carrier-2K complexed form.

The influence of E_m alone on the K fluxes in muscle at constant $[\mathrm{K}]_o$ and $[\mathrm{K}]_i$ can be determined in fibers in which E_m is chloride-clamped.[12] By changing $[\mathrm{Cl}]_o$ to vary E_m, both K efflux and influx can be measured at constant $E_\mathrm{K} = -18$ mV. When the potential is made more positive, both influx and efflux decline, K influx declining more rapidly with increasing $E_m - E_\mathrm{K}$ than K efflux. If the fluxes were following the laws for simple electrodiffusion, influx should decline but efflux should rise if P_K remains constant.

Under these conditions, however, P_K is probably declining,[3] so that it is difficult to make a test of the data. It is clear, however, that both variations in E_m alone and in $[K]_o$ alone influence the K fluxes in muscle. In some cases, the K fluxes can be kinetically described by the carrier model by considering variation of $[K]_o$ only and not E_m. Such cases include the steady-state unidirectional K fluxes and K efflux at a constant $E_m = -18$ mV. As the K fluxes under these conditions are adequately described by equation (60), it appears that the rate-limiting step for these fluxes involves the reaction of external K with the control site. At more positive values of the membrane potential ($E_m \geq -18$ mV), K influx declines very rapidly as $[K]_o$ approaches low values, indicating that the reaction of two K ions with the carrier is now becoming rate limiting. From this kind of an analysis, one can infer that the large internal negativity provided by the normal resting membrane potential of the fiber greatly aids the loading of the carrier from the outside. For steady-state conditions (solid points in Figure 4.1), E_m declines as $[K]_o$ is elevated. However, when E_m has reached depolarized values that would, of themselves, cause carrier loading with K to become rate limiting, $[K]_o$ has reached such high values that the effect is not observed. The upper curve in Figure 4.1 continues to be describable by equation (60) in which carrier activation is rate limiting. In the case where E_m is constant at $E_m = -18$ mV, the relative positivity of the potential renders it easy for the carrier to load with K ions from the inside of the fiber, and K efflux follows kinetics in which activation of the carrier by K_o is again rate limiting.

Though several facts can be rationalized by this approach, some important points remain unanswered. Notable among these are the facts that P_K falls to such low values when the driving force favors a large outward movement and that net K movement in muscle shows a strong rectification for inward currents.[2,3,14]

INTERACTIONS WITH FOREIGN CATIONS

The permeability of the muscle fiber membrane to Rb, Cs, and Tl ions has also been investigated.[4,14a] From the effects of external Rb and Cs on the resting membrane potential, P_{Rb} is about one-fourth and P_{Cs} about one-tenth the value of P_K. The permeability of muscle membrane to Tl ions is about the same as it is to K ions.[14a] The permeability coefficients for Rb and Cs deduced from the measurement of cation influx using tracers are lower than that for K, the rank order being $P_K > P_{Rb} > P_{Cs}$. There are quantitative discrepancies, however, between the P values deduced from E_m measurements and those deduced from influx measurements, particularly between P_K and P_{Rb}. The discrepancies could be due to a fraction of the influx occurring through pathways other than those responsible for permeability. Quantitative deviations would be expected, for example, if a fraction of K and Rb influx were due to active transport as first suggested by Adrian.[15]

Another finding with the foreign cations Rb and Cs is that their presence in the external medium depresses the value of P_K.[4] The slowing action of Rb and Cs on K movement was interpreted in terms of either a carrier model or a single file pore model by assuming that Rb and Cs ions compete with K ions for occupancy of the K carrier molecule or the K channel. A further study of the permeability of the muscle fiber membrane to K and to Rb was made by Adrian,[15] who found that Rb can substitute well for K ions in carrying outward currents through delayed-rectification channels during an action potential. Rubidium ions were ineffective, however, in being able to carry large inward currents over the pathway that normally permits K ions to carry large inward currents. The ability of Rb ions to suppress inward K movement and lower the permeability of the muscle fiber to K ions was also confirmed.

The kinetics of the process by which Rb and Cs ions reduce the rate of inward K movement suggest that these cations compete with K ions for carrier sites or for carrier activation sites.[13] If the rate of inward movement of K is determined by the fraction of sites occupied by K ions, and if foreign cations in the external medium compete with external K in a first-order reaction, equation (60) for K influx becomes altered as follows:

$$j_i = \bar{j}\left(\frac{[\mathrm{K}]_o}{[\mathrm{K}]_o + \dfrac{\mathrm{K}_m}{\mathrm{K}_I}[I]_o + \mathrm{K}_m}\right) \tag{61}$$

where I refers to the cation inhibiting K influx and K_I is the reciprocal of the affinity of the sites for the cation I. The order observed for the affinity of the carrier control sites for these cations is Rb > K > Cs. The rates of inward movement of Rb and Cs can also be fit by equations similar to equation (61), suggesting that the K carrier can also transport these cations.[13,16] The quantitative details of the inward transport of foreign cations must be revised, however, as it is now known that considerable active inward transport of Rb and Cs ions occurs into muscle fibers.[17,18]

Though foreign cations interfere with inward K movement when they are present in the external solution, their presence at similar concentrations inside of the muscle fibers does not interfere with the outward movement of K.[19] This fact suggests that the nature of the carrier reactions is quite different on the inner and outer membrane surfaces. It is interesting, however, that foreign cations present inside the fibers, but absent in the external medium, interfere with the inward movement of K.[16] The internal foreign cations are evidently able to reach the external control site from the inside to inhibit the activation by K_0. The alternative to this hypothesis is to assume that the foreign cations can occupy an internal membrane site to block influx of K but not the efflux of K. This explanation would demand that K influx and efflux occur over different carrier pathways. No doubt a carrier mechanism similar to that proposed by Horowicz et al.[12] could be modified to account for these observations.

The presence of foreign cations in the external medium reduces the rate of outward K movement as well as the rate of inward movement.[4,20] This finding is in line with the hypothesis of Horowicz et al.[12] that both K influx and efflux are mediated by the same carrier process and that the fundamental event underlying these K movements is the reaction of external K ions with a carrier control site in a first-order reaction. The presence of external foreign cations would be expected to reduce both K influx and efflux if they reacted with either the activation or control site of the carrier sites themselves in competition with K ions.

The activating effect of external K ions on K efflux is illustrated in Figure 4.2 for different external media.[21] In all cases, the Na in Ringer's solution has been replaced with an osmotic equivalent of the cation indicated on the figure. Replacement of Na with either tetramethylammonium ions (TMA) or with Li ions has little effect on activation of the carrier by $[K]_o$, whereas activation is considerably reduced in Tris or Mg Ringer's solution and abolished completely in Cs Ringer's solution. Under K-free conditions in the external medium, a residual activation is observed in all Ringer's solutions except Mg and Cs Ringer's solutions. This behavior can be interpreted in two ways: a small value of $[K]_o$ may exist under nominally K-free conditions due to K leaking from the fibers, or Na and the substituting cations themselves, in some instances, may be able to activate the carrier. That cations other than K can activate the K carrier in muscle fibers is clearly shown in the work of Beaugé et al.,[20] where it was observed that low concentrations of Cs ions in Mg-substituted Ringer's solution increase the K efflux in the absence of external K. The results shown in Figure 4.2 indicate clearly that the activating effect of small values of $[Cs]_o$ is completely abolished by raising the value of

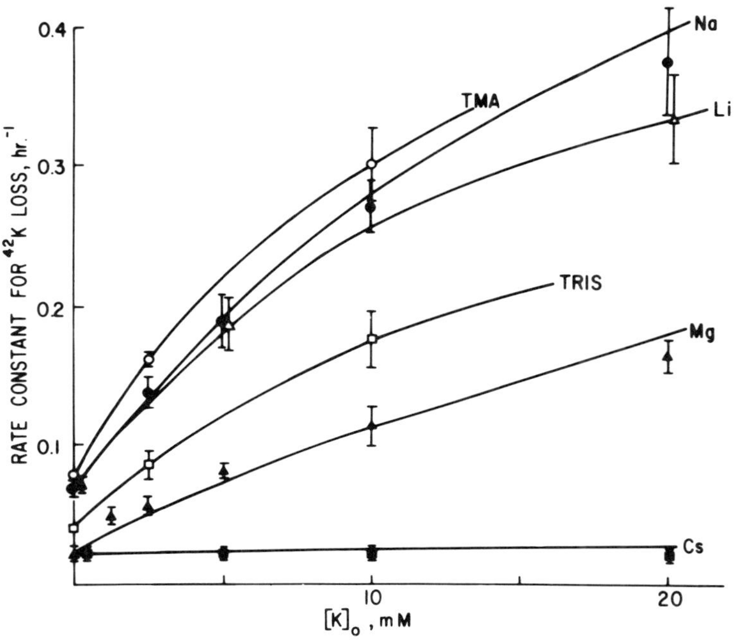

Figure 4.2. The rate of ^{42}K loss from frog sartorius muscle is plotted versus $[K]_o$ with different cations (see curves) used to replace Na ions in Ringer's solution. From R. A. Sjodin and S. C. Wu, unpublished results.

$[Cs]_o$ to around 100 mM, as the curves in Mg and Cs Ringer's solutions meet at exactly the same point at $[K]_o = 0$. It is evident that, regardless of whether or not other cations activate the K carrier, they can considerably modify the kinetics by which the K ions themselves activate the carrier.

Probably not enough is yet known about the K carrier in muscle fibers to further interpret the curves in Figure 4.2. It is noteworthy, however, that a divalent cation such as Mg can substantially modify the activation of the carrier by K. The kinetics of inhibition observed with lower concentrations of Mg and Cs are illustrated in Figure 4.3, where the rate constant for ^{42}K loss is plotted against concentration of the inhibiting cation. Also included in the figure are data obtained with external Ba ions. Clearly the affinity for Ba ions of the site at which inhibition of activation by $[K]_o$ occurs is much higher than its affinity for the other cations. The presence of Ba ions externally at concentrations of a few millimolar suffices to completely block the K_o-acti-

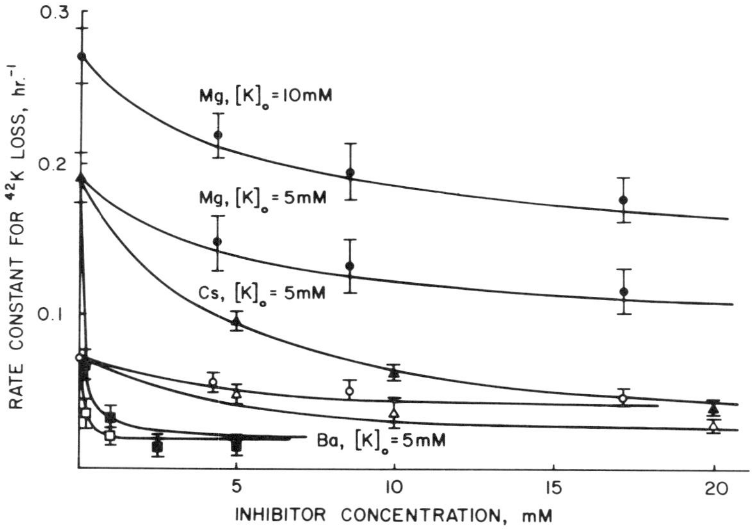

Figure 4.3. The rate of K loss using ^{42}K ions is plotted against the concentration of inhibiting cation for frog sartorius muscle. From R. A. Sjodin and S. C. Wu, unpublished results.

vated K efflux. The value of K efflux in $[\mathrm{Ba}]_o = 2.5$ mM solutions is less than 10 percent of the normal value of K efflux. The same low basal level of K efflux is obtained in mM Ba solutions, high concentrations of Cs, and in Mg Ringer's solution when $[\mathrm{K}]_o = 0$. It seems likely that this amount of efflux occurs via a mechanism that is not activated by external K ions. In frog sartorius muscle, the carrier-mediated K flux is around 7 pmoles/cm^2·sec when $[\mathrm{K}]_o = 2.5$ mM and around 12 pmoles/cm^2·sec when $[\mathrm{K}]_o = 5$ mM.

THE RELATION BETWEEN CARRIER FLUX OF POTASSIUM IONS AND MEMBRANE CONDUCTANCE TO POTASSIUM

A type of carrier process that requires that ions be transported into and out of the cell on a one-for-one basis cannot produce electrical current and, hence, cannot make a contribution to membrane conductance. There are reasons for supposing that the K carrier in muscle fibers is not of this type. It has already been pointed out that the isotopically determined fluxes of K in frog muscle are too low to account for the K conductances inferred from electrical measurements. If a large portion of the K fluxes were due to a 1:1 carrier or exchange diffusion, the fluxes measured with ^{42}K would be expected to give a calculated K conductance much higher than the electrically measured value.[15] Another reason for believing that the K carrier in muscle fibers can produce an electrical K current is the fact that Ba, the same cation that blocks most of the unidirectional K fluxes, also reduces the electrically determined total membrane conductance.[8] The total membrane conductance measured in normal Ringer's solution was 263 μmhos/cm^2, and this value fell to 196 μmhos/cm^2 in the presence of Ba ions. The value of the membrane conductance measured with Ba added agrees fairly well with the chloride conductance measured by Hodgkin and Horowicz.[3] The Ba-sensitive membrane conductance was 67 μmhos/cm.[2] It is difficult to make an exact comparison of this with values of G_K determined electrically, due to different experimental conditions. However, if only the data are used for which $[K]_o$ is nearly the same (2.5 mM), the average value of G_K inferred from the electrical measurements in the absence of Ba ions is 69.8 μmhos/cm^2.[3] It is evident that, to within the experimental errors, Ba ions block all of the membrane conductance to K in muscle fibers from the frog. The fact that Ba ions block essentially all of the K conductance and also the major part of the unidirectional K fluxes in muscle fibers provides powerful support for the notion that the carrier responsible for the unidirectional K fluxes also provides the membrane conductance to K ions.

The structural details of the K carrier remain unknown. Data presently available do not permit one to distinguish between a mobile carrier,[12] an immobile carrier,[16,22] and an aqueous gated channel.[8,10] It has been proposed that K ions may cross the cell membrane by having a part of their hydration replaced by polar groups in membrane channels.[23] Barium ions have nearly the same crystal radii as K ions and would therefore be expected to fit the K channel well if the entering entity is the ion with a single shell of water molecules. Being doubly charged, Ba ions would be bound more tightly to negative sites within the channel, thus accounting for the intense blocking action. On the other hand, the tight binding of a Ba ion at a carrier control site would prevent K ions from activating the carrier. Such models may be helpful in suggesting further experimentation.

This monograph deals with ion transport in resting muscle fibers. It does not, therefore, deal with ion flows that occur during action potentials and during voltage clamps. The ion fluxes dealt with here occur over long periods of time such as several minutes or even hours. There are studies, however, in which electrical currents are passed across muscle fiber membranes for periods of the order of one second to determine membrane conductances. As already is evident, these studies may be fairly closely related to those in which ionic fluxes are measured over much longer periods of time.

For inward currents across the muscle fiber membrane, the conductance is higher than for outward currents. This rectification is due to K ion movement since it occurs in Cl-free media.[14] As this kind of rectification is opposite to the delayed rectification responsible for large outward currents during the action potential, it has been termed "anomalous rectification." Further studies of this phenomenon have indicated that Rb ions[15] and Ba ions[8] reduce the anomalous rectification due to K ions. As two cations that inhibit anomalous rectification also inhibit the membrane K carrier for the unidirectional K fluxes, it is tempting to conjecture that the two phenomena share the same mechanism. According to this view, the membrane K carrier that produces equal and opposite K unidirectional fluxes in the steady state can become unbalanced to produce a net K current. The effects of membrane potential on such a carrier would be complex and asymmetrical. The carrier would become unbalanced more easily for hyperpolarizing currents than for depolarizing currents. In addition, the apparent P_K would show a continuous decline, with increasing positivity within the fibers relative to E_K.[3,12,24]

A complicating factor in attempting such correlations, however, is the time dependence of the conductance change underlying anomalous rectification.[24] The kinetics of this process suggest that there are two pathways for K movement across the muscle fiber membrane, one located in the surface membrane and another in the transverse tubules of the T-system. The exact location of the membrane K carrier is unknown, but the kinetics of the isotopic K fluxes are consistent with a location in the surface membrane.

SUMMARY

Though the detailed mechanisms involved in the electrodiffusion of K across the muscle fiber membrane are largely unknown at present, several statements can be made that characterize this kind of K transport in muscle.

1. When K conductance is referred to nearly the same value of $E_m - E_K$, G_K and unidirectional K fluxes increase with increasing $[K]_o$ in roughly the same proportion.

2. The electrical membrane conductance to K near the steady state is about three times the value of G_K calculated from unidirectional fluxes on the basis of the independence principle.
3. The value of P_K determined from near-steady-state unidirectional K fluxes is around 1×10^{-6} cm/sec and remains reasonably independent of $[K]_o$ and E_m over a wide range in these variables. The value of P_K determined from electrical measurements near the steady state is $1\text{-}2 \times 10^{-6}$ cm/sec, in fair agreement with the value calculated from unidirectional fluxes. The value of P_K determined from K currents, however, shows rectification and rises for inward K movement and falls for outward movement. The highest observed values are around 7×10^{-6} cm/sec and the lowest around 0.04×10^{-6} cm/sec.
4. Unidirectional K efflux measured with ^{42}K ions rises with increasing $[K]_o$ in a manner suggesting that the flux is carrier mediated and activated by $[K]_o$ at a single externally directed control site. This conclusion holds for all cases tested, whether E_m varies with $[K]_o$ or whether E_m is held constant at -18 mV as $[K]_o$ is varied.
5. Unidirectional K influx near-steady-state conditions also follows first-order carrier kinetics as though the rate-limiting step were activation of the carrier by K at a single externally directed site. When E_m is held constant at a value of -18 mV, however, K influx increases with rising $[K]_o$ in a manner indicating that the rate-limiting step is now the reaction of two K ions with the carrier. These differences in behavior can be reconciled if it is supposed that carrier loading with K and carrier activation by K are two separate processes, each of which can be rate limiting under different circumstances.
6. The unidirectional K fluxes approximately obey the flux ratio equation for independent ion movements when the K fluxes are near the steady state and when the fluxes are not in the steady state but $[K]_o$ is at a low value ($\cong$ 0.2 mM) and $E_m \cong -110$ mV. For non-steady-state conditions with reference to an E_m value of -18 mV, the K flux ratio does not obey the independence principle but follows the square of the value calculated on the basis of independence.
7. Foreign cations inhibit unidirectional K fluxes when present externally and also reduce P_K determined electrically. Cations that are particularly effective are Rb, Cs, and Ba.
8. Barium ions strongly inhibit unidirectional carrier-mediated K fluxes in muscle fibers when present externally. They also abolish anomalous

rectification. It is tempting to conclude that the two processes occur by way of a common mechanism.

9. The K fluxes in muscle fibers are basically nonindependent in nature. The evidence for this is that conductances calculated from fluxes do not agree with measured conductances, the flux ratio for K does not always obey the equation based on independence, and foreign cations interact strongly with the K fluxes. The conclusion that K efflux and influx occur over a pathway with different properties that depend upon the direction of transit is supported by the fact that external Rb strongly inhibits K influx but internal Rb has little effect on K efflux.[19]

10. The same carrier mechanism that is responsible for the unidirectional K fluxes measured with ^{42}K ions is also probably responsible for net K ionic currents, which it could produce by simply becoming imbalanced and transporting K inwardly and outwardly at different rates. The detailed mechanism for how the membrane potential influences the carrier and how $[K]_o$, $[K]_i$, and E_m interact to unbalance the carrier is very complex.[12]

The remaining mechanism of K ion movement in muscle fibers is active transport, which is covered in Chapter 6.

REFERENCES

1. R. H. Adrian, *J. Physiol.* **143,** 59P (1958).
2. R. H. Adrian, *J. Physiol.* **151,** 154 (1960).
3. A. L. Hodgkin and P. Horowicz, *J. Physiol.* **148,** 127 (1959).
4. R. A. Sjodin, *J. Gen. Physiol.* **42,** 983 (1959).
5. R. A. Sjodin and E. G. Henderson, *J. Gen. Physiol.* **47,** 605 (1964).
6. R. D. Keynes, *J. Physiol.* **114,** 119 (1951).
7. P. Fatt and B. Katz, *J. Physiol.* **115,** 320 (1951).
8. N. Sperelakis, M. F. Schneider, and E. J. Harris, *J. Gen. Physiol.* **50,** 1565 (1967).
9. A. L. Hodgkin and R. D. Keynes, *J. Physiol.* **128,** 61 (1955).
10. R. A. Sjodin, *J. Gen. Physiol.* **48,** 777 (1965).
11. R. A. Sjodin and L. A. Beaugé, *J. Gen. Physiol.* **52,** 389 (1968).

12. P. Horowicz, P. W. Gage, and R. S. Eisenberg, *J. Gen. Physiol.* **51,** 193S (1968).

13. R. A. Sjodin, *J. Gen. Physiol.* **44,** 929 (1961).

14. B. Katz, *Arch. Sci. Physiol.* **3,** 285 (1949).

14a. L. J. Mullins and R. D. Moore, *J. Gen. Physiol.* **43,** 759 (1960).

15. R. H. Adrian, *J. Physiol.* **175,** 134 (1964).

16. V. Bolingbroke, E. J. Harris, and R. A. Sjodin, *J. Physiol.* **157,** 289 (1961).

17. R. H. Adrian and C. L. Slayman, *J. Physiol.* **184,** 970 (1966).

18. L. A. Beaugé and R. A. Sjodin, *J. Physiol.* **194,** 105 (1968).

19. R. A. Sjodin, *J. Cell. Comp. Physiol.* **66,** Suppl. 2, 27 (1965).

20. L. A. Beaugé, A. Medici, and R. A. Sjodin, *J. Physiol.* **228,** 1 (1973).

21. R. A. Sjodin and S. Wu, unpublished results (1972).

22. E. J. Harris and R. A. Sjodin, *J. Physiol.* **155,** 221 (1961).

23. L. J. Mullins, *Ann. N.Y. Acad. Sci.* **94,** 390 (1961).

24. R. H. Adrian and W. H. Freygang, *J. Physiol.* **163,** 61 (1962).

5

Mechanisms of Sodium Ion Movement in Muscle

From the previous discussions of the membrane potential and its relation to E_K, P_{Na} is of the order of 0.01 P_K. To deduce the permeability coefficient from ion flux measurements is much more difficult for Na than for K, due to the complicating presence of other flux components. In the case of the K fluxes, there is no evidence for exchange diffusion, and as a first approximation, active K inward transport in normal muscle fibers is low enough to be neglected in the determination of passive fluxes. In muscle fibers with normally low values of $[Na]_i$, at least 95 percent of K influx is passive. The situation is quite different for Na ions, where almost all of the Na efflux is a combination of active transport and exchange diffusion with half the Na influx being due to exchange diffusion.

MEMBRANE PERMEABILITY TO SODIUM IONS

A component of Na flux that is due solely to a leakage through passive Na channels can be determined by measuring unidirectional Na influx with a radioactive tracer at various values of $[Na]_i$ and extrapolating the data to $[Na]_i = 0$.[1] Under these conditions, both Na exchange diffusion and Na active transport become zero in the limit, as both flux components require some Na_i for their occurrence. The value of Na influx obtained in this way is 2.7 pmoles/cm^2·sec. As $[Na]_i = 0$, equation (55) can be used to calculate P_{Na} with $E_m = -92$ mV and $[Na]_o = 120$ mM. The value obtained for P_{Na} is 0.6×10^{-8} cm/sec, which is very close to 0.01 times the value of P_K obtained from K unidirectional flux data.

Very nearly the same value of P_{Na} is obtained if net Na influx is measured in a K-free medium in the presence of ouabain.[1] Under these conditions, Na pumping should not occur and exchange diffusion will not be included, as the net change in $[Na]_i$ is measured flame photometrically and one-for-one exchanges are not included. The value of P_{Na} determined in this way is independent of both $[Na]_i$ and $[Na]_o$. Net Na influx is constant with time over a period of several hours in which $[Na]_i$ changes severalfold. Also, the net Na influx obeys equation (55), as the value of the flux increases linearly with increasing values of $[Na]_o$. Increasing the value of $[K]_o$ decreases the value of the net Na influx. Much of the change in net Na influx can be attributed to the depolarization induced by $[K]_o$ in accordance with equation (55). In addition, rising values of $[K]_o$ may result in slight reductions in the magnitude of P_{Na}.

To summarize, resting values of P_{Na} are reasonably constant, independent of both concentration changes and changes in E_m. In addition, the passive Na influx occurs in accordance with electrochemical flux theory summarized in equations (9), (54), and (55). The use of flux ratio theory (equation 10) to calculate the value of passive Na efflux from the value of passive Na influx results in values of Na efflux that are too small to be separated experimentally from the other components of Na flux. One would not, therefore, expect to be able to measure the passive Na efflux under normal circumstances.

UNIDIRECTIONAL SODIUM FLUXES

As just explained, the measurement of unidirectional Na influx provides accurate information about P_{Na} only in the limit as $[Na]_i \rightarrow 0$. As $[Na]_i$ is raised to levels where about half of the muscle K has been replaced by Na, the Na influx rises linearly.[1] The empirical equation for unidirectional Na influx that fits the data is

$$J_i = j_{i0} + k_{\mathrm{Na}}[\mathrm{Na}]_i \tag{62}$$

where J_i is total unidirectional Na influx in units of pmoles/cm^2·sec, j_{i0} is the passive electrodiffusional inward leakage rate of Na (2.7 pmoles/cm^2·sec), and k_{Na} is a constant conversion factor numerically equal to 0.24 when $[\mathrm{Na}]_i$ is expressed in units of mM/kg fiber water. When uptake is expressed in units of μmoles/g(fiber)·hr, j_{i0} = 4 μmoles/g·hr, $[\mathrm{Na}]_i$ is expressed in units of μmole/g, and k_{Na} is a rate constant equal to 0.56 hr^{-1}. By contrast, the net influx remains constant at about 2 pmoles/cm^2·sec over the entire range in $[\mathrm{Na}]_i$. The additional component of Na influx that is proportional to $[\mathrm{Na}]_i$ must, therefore, represent a one-for-one exchange of Na_o for Na_i. This behavior would be expected from the work of Keynes and Swan[2], where it was shown that in normal muscle fibers, about one-half the Na efflux was due to an exchange diffusion of Na. Exchange diffusion was demonstrated by showing that Na efflux measured with ^{24}Na was approximately halved when external Na was replaced with either Li or choline. If Li or choline cannot exchange with internal Na, the portion of Na efflux resulting from exchange with external Na ions will disappear upon removal of external Na. The ^{24}Na is lost from muscles according to a first-order process in Na Ringer's solution and in Li Ringer's solution, provided that the value of $[\mathrm{Na}]_i$ does not become too low. The behavior of Na efflux at very low values of $[\mathrm{Na}]_i$ is discussed in subsequent sections.

As indicated, the unidirectional Na influx measured with tracer Na rises linearly with increasing $[\mathrm{Na}]_i$ from a value at $[\mathrm{Na}]_i = 0$ corresponding to the inward electrodiffusion of Na. One might inquire about the way in which unidirectional Na influx varies with $[\mathrm{Na}]_o$. Unlike the dependence on $[\mathrm{Na}]_i$, the dependence on $[\mathrm{Na}]_o$ is not linear[2]; a distinct bend in the curve occurs at low values of $[\mathrm{Na}]_o$. The total influx curve is the sum of a linear flux component and a curve for exchange diffusion obtained from efflux data. The linear component is evidently due to electrodiffusion of Na. The relations described for unidirectional Na influx are summarized in Figures 5.1 and 5.2. The extrapolated value for Na influx in Figure 5.1 gives reasonable agreement with the value of net Na influx determined flame photometrically under K-free conditions and in the presence of ouabain to inhibit the Na pump.[1] The net Na influx is known accurately and has been shown to be independent of $[\mathrm{Na}]_i$.

In summary, the major characteristics of Na influx into muscle fibers are that the electrodiffusional component is independent of $[\mathrm{Na}]_i$ and varies linearly with $[\mathrm{Na}]_o$, and that the exchange diffusion component increases

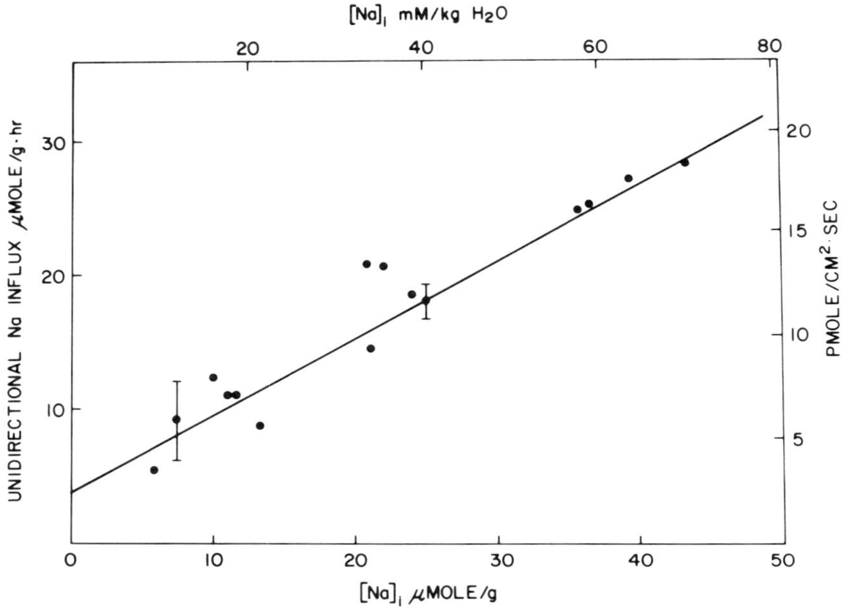

Figure 5.1. Sodium influx measured using ^{22}Na as tracer is plotted against the Na ion concentration in the muscle (frog sartorius). Points with vertical lines are averages of four to six experiments ± 1 S.E. From R. A. Sjodin and L. A. Beaugé, *J. Gen. Physiol.* **61,** 222 (1973).

linearly with $[Na]_i$ and nonlinearly with $[Na]_o$. Under some conditions, an additional component due to active Na transport becomes apparent. This component is discussed along with properties of the Na pump.

It is clear that Na efflux has at least three components: active Na transport, exchange diffusion, and electrodiffusion. For most practical cases, electrodiffusive Na efflux can be neglected because of its small numerical value. A first approach to analyzing Na efflux is thus to distinguish between active transport and exchange diffusion. This is accomplished via variations in $[Na]_i$, $[Na]_o$, $[K]_o$, and the addition of cardioactive steroids to the external medium.

The enhancing effect of increased $[K]_o$ on the efflux of tracer Na from frog muscle fibers was first noted by Harris.[3] The sensitivity of Na efflux to $[K]_o$ was further studied by Keynes,[4] who observed that changing the value of $[K]_o$ from 2.5 to 10 m*M* reversibly increased the Na efflux. On the other hand, removal of K from Ringer's solution reversibly decreased the Na efflux. The

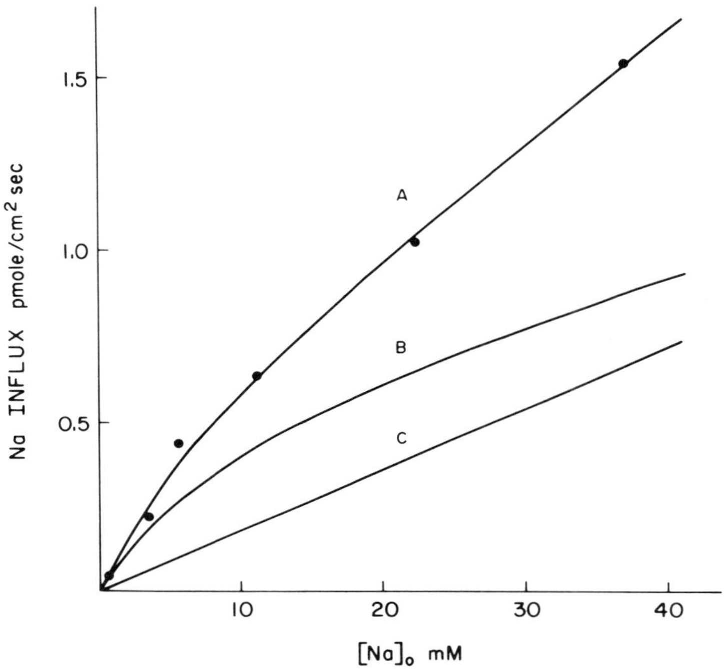

Figure 5.2. Sodium influx in frog sartorius muscle as measured with ^{24}Na is plotted as a function of $[Na]_o$. Total Na influx, curve *A*, is replotted from data presented in R. D. Keynes and R. C. Swan, *J. Physiol.* **147,** 591 (1959). Curves *B* and *C* give the "coupled" and linear components of Na influx deduced by these authors.

average values of the rate constants for loss of ^{24}Na from muscles were: $[K]_o = 0$ mM, 0.64 hr^{-1}; $[K]_o = 2.5$ mM, 1.12 hr^{-1}; and $[K]_o = 10$ mM, 1.53 hr^{-1}.

Keynes also noticed that the sensitivity of Na efflux to $[K]_o$ was increased when muscles were previously exposed to a K-free medium for a prolonged period.[4] As exposure of muscle fibers to K-free solutions results in K loss and Na gain,[5] this work suggests that $[Na]_i$ plays a role in determining the sensitivity of Na efflux to $[K]_o$, high values of $[Na]_i$ favoring an increased sensitivity to $[K]_o$. It was subsequently found that addition of g-strophanthidin (ouabain) to the medium at levels of 1–5 μg/ml inhibited Na efflux in much the same way as omission of K from the medium.[6] Ouabain abolished the K-sensitive part of the Na efflux. The same kind of observation was made using small bundles of frog semitendinosus muscle, where a substantial fraction of the Na efflux was abolished by external application of 10^{-5} M strophanthidin.[7]

The Na efflux in muscle fibers is thus sensitive to both $[K]_o$ and $[Na]_o$. Muscles enriched with Na by prolonged exposure to K-free Ringer's solution recover their normal low Na content when K ions are readmitted to the medium.[5,8-10] The conditions for recovery of the fibers to a low Na content are adequate external K and sufficiently high ambient temperature. The Na extrusion under these conditions takes place against an electrochemical gradient and requires metabolic energy. Such an extrusion is clearly the result of a Na pump. This Na extrusion is abolished by the external application of cardioactive steroids such as strophanthidin or ouabain.[11,12] As both net Na extrusion against an electrochemical gradient and tracer Na efflux are inhibited by removing external K and by applying cardioactive steroids, it seems reasonable that both processes result from the same mechanism, namely, the operation of a metabolically mediated Na pump. In view of this correlation, the K-sensitive and ouabain-sensitive fraction of Na tracer efflux is taken to be a measure of Na pump activity. It becomes evident in a later section that this component of Na efflux, though highly significant, does not reflect the total Na pump activity.

As originally envisioned, exchange diffusion and active transport represent two distinct processes. If external Na ions did not influence the Na pump and if external K ions did not affect exchange diffusion, one would expect the Na-sensitive and K-sensitive portions of Na efflux to be independent and additive. Actually, however, they were not independently additive.[2] The simultaneous removal of K and most of the external Na, for example, resulted in an efflux reduction that was only 85 percent of the sum of the flux reductions occurring when K and Na ions were removed separately. In addition, the K-free and Na-free effects tended to vary in a reciprocal manner. As muscles aged and gained Na, the fraction of the Na efflux that was sensitive to K increased, while the fraction that was sensitive to Na decreased. At an internal Na concentration of about 20 mM, the sensitivity of Na efflux to external Na was lost completely if Li was the cation used to replace the Na. However, a partial restoration of the initial K and Na sensitivities was obtainable if the aged muscles were allowed to soak in an Na-free medium for a prolonged period of time.[2] These findings suggest that the value of $[Na]_i$ plays a role in determining the sensitivity of Na efflux to external Na and K, and that exchange diffusion and active transport have more complex external cation dependencies than simple additive processes would have.

In frog sartorius muscles that have been specially treated to reduce their Na content by one-half, the fraction of Na efflux sensitive to external Na rises to about 80 percent.[13] For similarly treated muscles, 10^{-5} M strophanthidin reduced the Na efflux by only about 25 percent. If Li ions are used to replace Na ions, the addition of strophanthidin to Li Ringer's solution approximately halves the remaining Na efflux to give a residual Na efflux equal to about 10 percent of the total Na efflux. The nature of the residual Na efflux is largely unknown, but it probably represents a cardioactive steroid-insensitive Na active transport uncoupled to either external Na or K, as its magnitude is some 30 times greater than the magnitude of the electrodiffusional efflux of Na. For muscle fibers with low $[Na]_i$, almost all of the Na efflux is a combination of Na-sensitive and strophanthidin-sensitive components; in this case, these components are independently additive.

In the case of small bundles of fibers isolated from the frog's semitendinosus muscle, these figures are approximately reversed; 80 percent of the Na efflux is abolished by strophanthidin, leaving only a minor fraction due to Na:Na exchange diffusion.[7] Evidently, great variation in the quantitative fractions of Na efflux due to active transport and exchange can be expected in muscle fibers; this may be due to different fiber types. Some of the differences may also be attributed to varying values of $[Na]_i$ under different experimental conditions. In sartorius muscle fibers with elevated values of $[Na]_i$, about 60 percent of the Na efflux is abolished by external application of 10^{-5} M strophanthidin.[13]

The dependence on $[Na]_i$ and the interrelationships between Na-sensitive and K-sensitive Na fluxes in muscle fibers become understandable with a thorough knowledge of the cation-activation kinetics of the Na pump. A powerful clue is provided by the apparent behavior of exchange diffusion at elevated $[Na]_i$. It was stated that the dependence of Na efflux on $[Na]_o$ was lost when $[Na]_i$ was raised to around 20 mM. If, however, the value of $[Na]_i$ is increased still further, the dependence of Na efflux on external Na, with Li as an Na replacement, returns but the effect is reversed; Na efflux is higher in the Li solution than in the Na solution.[14,15] The reversed exchange diffusion effect in muscle fibers with elevated $[Na]_i$ is abolished by external application of either ouabain or strophanthidin (10^{-5} M) and is converted to the usual Na-free effect. The reversed exchange diffusion effect in fibers with elevated $[Na]_i$ does not occur if tris rather than Li ions are used to replace Na even in the absence of cardioactive steroids.[16] When tris is used to replace $[Na]_o$, only reductions in Na efflux are observed whether $[Na]_i$ is normal, low, or elevated. The reversed Na-free effect observed with Li at elevated $[Na]_i$ would be expected if Li ions could stimulate the Na pump.[15,17,18] The dependence of the Na-free effects on $[Na]_i$ observed by Keynes and Swan[2] would also be explained. This subject is discussed further, along with the modes of activation of the Na pump in muscle fibers.

In summary, Na influx is the sum of electrodiffusional Na leakage and an exchange component that increases linearly with increasing $[Na]_i$ and nonlinearly with increasing $[Na]_o$. The efflux of Na consists of a K-activated Na pump inhibited by cardioactive steroids, an exchange diffusion component, an uncoupled residual ouabain-insensitive active transport, and a numerically insignificant outward electrodiffusional leak. The exchange component of Na efflux varies linearly with $[Na]_i$ over a considerable range, and nonlinearly with $[Na]_o$, showing saturation at Ringer's solution concentrations of $[Na]_o$. The influx and efflux components of Na exchange share the same kinetics and apparently reflect a one-for-one Na : Na exchange.

The influx of Na is normally insensitive to cardioactive steroids; this suggests that exchange diffusion may be a distinct mechanism apart from the Na pump.[7] Exchange diffusion could also be due to a part of the Na pump machinery that happens to be uninfluenced by cardioactive steroids. The exchange diffusion and active transport fractions of Na efflux vary in reciprocal proportion as $[Na]_i$ varies; high $[Na]_i$ favors a high proportion of active transport and a low proportion of exchange diffusion, the reverse being true in the case of a lower than normal $[Na]_i$. When the value of $[Na]_i$ is normal or lower than normal, exchange diffusion and active transport of Na behave in an independent, additive manner. This suggests either separate mechanisms or fairly distinct parts of the same mechanism. In making such determinations it is important to replace Na with a cation that does not itself affect the Na pump. As the fractions of efflux due to exchange diffusion and to active transport vary in a reciprocal manner, it is tempting to postulate a common mechanism with a limited total capacity and interconvertible modes of operation depending upon the value of $[Na]_i$. Such considerations must remain speculative, however, as there are other explanations for exchange diffusion.[19]

THE SODIUM PUMP IN MUSCLE FIBERS

It has been decisively demonstrated in red blood cells[20-22] and in squid giant axons[23] that ATP is the energy source for the extrusion of Na ions by the Na pump. As the Na pump in muscle fibers has much in common with the Na pump in red blood cells and nerve fibers, it was to be expected that ATP also is the substrate for active transport in muscle.[24] It has long been known that active Na transport is closely related to an $(Na^+ + K^+)$-stimulated ATPase activity located in cell membranes.[25-27]

A scheme has emerged in which the $(Na^+ + K^+)$-stimulated ATPase is an integral part of the Na pump and binds Na and ATP at the inner membrane surface. Through the formation of a phosphorylated intermediate, Na ions are transported to the outer membrane surface, where K ions are bound. This K binding induces hydrolysis of ATP, release of Na ions to the outside, and inward transport of K ions. In the simplest version of this scheme, the K ions that activate at the outer surface of the membrane are also the same ones that are actively transported inward. In this case, K-activation and K active transport should display similar kinetics. This need not always be the case, however, as shown by kinetic studies in squid giant axons.[28] More complicated kinetics are also suggested by studies of isolated $(Na^+ + K^+)$-stimulated ATPase preparations where multiple Na and K binding sites have been demonstrated.[29,30] As both Na and K ions are actively transported by the pump under discussion, active Na and K movements do not occur independently and must, at some point, be discussed as part of the general subject of Na:K pump stoichiometry. As special problems are encountered in studying active K transport in muscle fibers, the K part of the pump and the stoichiometry of the Na:K pump are dealt with in Chapter 6.

ACTIVATION OF THE SODIUM PUMP BY EXTERNAL POTASSIUM

It has already been stated that Na efflux in muscle fibers is sensitive to $[K]_o$ and that the magnitude of the sensitivity is highly influenced by the value of $[Na]_i$. The K-sensitive Na efflux in muscle fibers is also highly dependent on the value of $[Na]_o$.[16] For this reason, the value of $[Na]_o$ must be specified in studies of K activation of the pump. Removal of Na from Ringer's solution and replacement with tris results in an increased component of K-sensitive Na efflux. In a Na-free medium, the K-sensitive Na efflux rises with increasing $[K]_o$ along a rectangular hyperbolic curve that shows saturation at high values of $[K]_o$. This component of Na efflux can be fit by a relation of the form

$$m_{Na} = M_{Na}\left(\frac{[K]_o}{[K]_o + k_{1/2}}\right) \tag{63}$$

where m_{Na} is the K-sensitive Na efflux and M_{Na} is the maximal value of this component.[16] The value of M_{Na} is linearly dependent on $[Na]_i$ provided $[Na]_i$ is not allowed to fall much below the normal value. The value of $k_{1/2}$ required to fit the data is 3.3 mM, and at the normal value of $[Na]_i$, M_{Na} is around 10 pmoles/cm^2·sec. When half the internal K has been replaced with Na, the value of M_{Na} increases to 65 pmoles/cm^2·sec. When equation (63) is written in reciprocal form, a plot of m_{Na}^{-1} against $[K]_o^{-1}$ yields a straight line:

$$\frac{1}{m_{Na}} = \frac{1}{M_{Na}} + \frac{k_{1/2}}{M_{Na}}\left(\frac{1}{[K]_o}\right) \tag{64}$$

A plot of the data for activation of the Na pump in muscle by $[K]_o$ in Na-free media is illustrated in Figure 5.3 in reciprocal form. The kinetic relations stated in equations (63) and (64) are similar to those for the velocity of an enzymatic reaction as it is affected by the concentration of substrate. In the case where $[Na]_o = 0$, K is analogous to a substrate that reacts with a single enzymatic site to produce the reaction. As the Na pump is enzymatic in character, the analogy is a good one. In the transport case, the products are K ions translocated to the inside, Na ions translocated to the outside, and the products of hydrolysis of ATP. The reaction with K ions is only one of several partial reactions of the pump,[31,32] but is rate limiting under the conditions of these measurements.

The foregoing results apply to studies of whole sartorius muscle. Due to the presence of a significant extracellular space and possible ion leaks from the muscle fibers, it is difficult to obtain the degree of stirring required to substantially reduce extracellular concentrations just outside the fiber membranes. It is possible, therefore, that K-free conditions in the bathing medium do not ensure K-free conditions just outside the fibers. Should this be the case, the basal level of Na efflux in solutions with $[K]_o = 0$ would already include a degree of activation by K that would make precise determinations of the true value of $k_{1/2}$ impossible. That difficulties of this nature may be present in the whole muscle studies is suggested by the fact that the whole

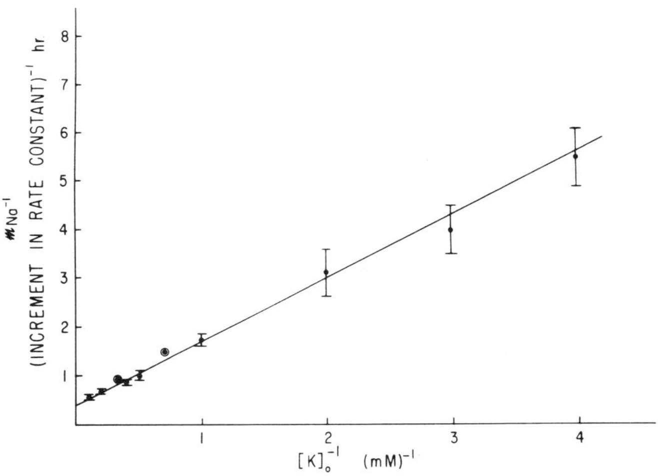

Figure 5.3. The reciprocal of the K-sensitive portion of the rate constant for ^{22}Na loss in the absence of external Na ions is plotted against the reciprocal of the external K ion concentration for frog sartorius muscle. External Na is replaced by an osmotic equivalent of tris. Bars refer to averages ± S.E. for six to nine measurements; circles refer to averages of two measurements each. From R. A. Sjodin, *J. Gen. Physiol.* **57,** 164 (1971).

muscle $k_{1/2}$ for $[Na]_o = 0$ is 3.3 m*M*, which is some eight times larger than the $k_{1/2}$ found in intact erythrocytes under similar conditions.[26] Similar studies of single fibers and small bundles of fibers from the frog's semitendinosus muscle yield a value of $k_{1/2}$ of 0.3–0.4 m*M*.[33]

The problem of an unstirred layer that accumulates K at low values of $[K]_o$ is not nearly as serious in single muscle fibers. As single muscle fiber and red cell values of $k_{1/2}$ agree rather well, it is evident that whole sartorius muscle may present quantitative difficulties at very low external ion concentrations. It should be emphasized that the differences resulting from using whole muscle are quantitative only and not qualitative, as the K-activation data in single fibers can be adequately fit by equations (63) and (64). It would obviously be advantageous to employ single muscle fibers in ion transport studies whenever possible. Whole muscle studies are necessary, however, as the difficulties of single-fiber flux work preclude a large quantity of experimentation. If one had to rely only on single-fiber work for information on ion transport in muscle, much less would be known than is now known.

THE INFLUENCE OF $[\mathrm{Na}]_o$ ON ACTIVATION OF THE SODIUM PUMP BY POTASSIUM

The effect of external Na ions on pump activation by K must be determined against a complicated background of possible responses of Na efflux to changes in $[\mathrm{Na}]_o$ due to both exchange diffusion and pump effects. To remove the effects of exchange diffusion as much as possible, use is made of muscles in which the Na content has been artificially elevated. Muscles with various elevated Na contents can be prepared simply by storing them at low temperatures in K-free Ringer's solution for varying periods of time. In muscles with an elevated Na content, the K- and ouabain-sensitive fraction of Na efflux increases and that fraction due to exchange diffusion decreases. Such muscles, in the presence of external K ions, show a decrease in Na efflux when Na ions are added to a previously Na-free, tris-substituted medium.[16] This is opposite to the kind of response observed in fresh muscles with a normal Na content. By isolating the fraction of Na efflux responsible for the flux reduction seen with increasing values of $[\mathrm{Na}]_o$ in Na-enriched muscles, it has been shown that external Na has an inhibitory action on the K-activated part of the Na pump.[16]

Increased concentrations of external Na change the kinetics of K activation from a Michaelis-Menten type to a type showing a marked sigmoidal character. Curves that illustrate the influence of $[\mathrm{Na}]_o$ on pump activation by $[\mathrm{K}]_o$ are shown in Figure 5.4. This behavior is similar to that observed for the Na pump in red blood cells[26] and in squid giant axons.[34] As the pump rate measured at very high values of $[\mathrm{K}]_o$ seems to approach the same maximal rate, competitive inhibition by $[\mathrm{Na}]_o$ is suggested. If external Na ions occupy the sites at which K ions activate, a reduction in activation by K would be expected. The sigmoidal nature of the curves suggests a multiplicity of activation sites. The kinetics observed in muscle can be described by a model in which there are three externally directed activation sites that must be simultaneously occupied by K ions in order for activation to occur. The presence of Na ions at one or more of the different sites prevents activation from occurring. The equation that describes this phenomenon is[16]

$$m_{\mathrm{Na}} = \frac{M_{\mathrm{Na}}}{1 + (k_{1/2} + a[\mathrm{Na}]_o)/[\mathrm{K}]_o + b([\mathrm{Na}]_o/[\mathrm{K}]_o)^2 + c([\mathrm{Na}]_o/[\mathrm{K}]_o)^3} \qquad (65)$$

where other symbols are as before and a, b, and c are ratios of rate constants and their products.

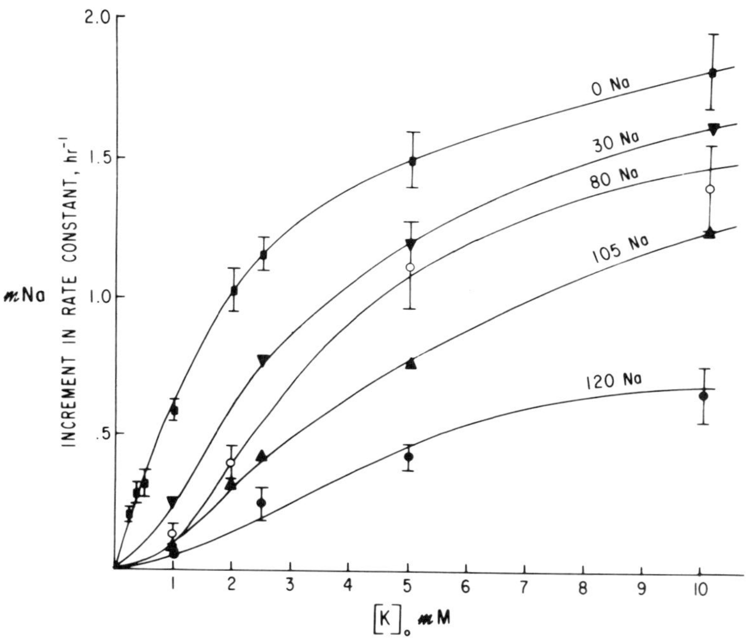

Figure 5.4. The K-sensitive portion of the rate constant for ^{22}Na loss from Na- enriched frog sartorius muscle is plotted versus $[K]_o$ for different values of $[Na]_o$. Numbers above curves refer to m*M* concentrations of Na, tris being used to balance osmolarity to 120 m*M* Na. Bars are ± 1 S.E. for six to nine measurements. From R. A. Sjodin, *J. Gen. Physiol.* **57**, 164 (1971).

A different explanation for the observed kinetics would be to suppose that external Na ions modify the affinity of a single activation site for K ions according to the allosteric enzyme activation model. Similar studies have been performed on fragmented membrane $(Na^+ + K^+)$-ATPase systems but conditions analogous to those under which transport occurs are difficult to match because both inner and outer enzymatic sites are exposed to the same medium. However, under appropriate conditions, Na and K ions have both stimulatory and inhibitory effects on enzymatic activity. Also, sigmoidal kinetics are observed when ionic concentrations are varied.[25] In erythrocyte ghost studies, $(Na^+ + K^+)$-ATPase activity can be measured with the directional characteristics maintained intact.[26,27] In these experiments, high values of $[Na]_o$ resulted in a decrease in the apparent affinity for K_o as observed in transport studies in red blood cells, squid giant axons, and skeletal muscle fibers.

ACTIVATION BY $[Na]_o$ IN THE ABSENCE OF EXTERNAL POTASSIUM

It is established that K-sensitivity and ouabain-sensitivity of Na efflux in muscle fibers are similar and that both fractions of Na efflux may be taken to represent the Na:K coupled part of the Na pump activity. In a more detailed analysis of flux components, however, it was found that K-sensitive Na efflux and ouabain-sensitive Na efflux were significantly different in magnitude.[17] The ouabain-sensitive Na efflux is greater than the K-sensitive Na efflux. When $[K]_o = 5mM$, ouabain-sensitive Na efflux exceeds K-sensitive Na efflux by about 20 percent. As the value of $[K]_o$ increases, the percentage difference between these two measured quantities decreases.

The difference between K-sensitive and ouabain-sensitive Na efflux would have an explanation if, in the absence of external K, Na ions themselves were able to provide some activation at external sites. A study of Na efflux as a function of $[Na]_o$ in K-free media with and without cardioactive steroid provides support for this view.[16] A summary of this study appears in Figure 5.5. Here it can be observed that a strophanthidin-sensitive component of Na efflux occurs under K-free conditions. This component increases, as $[Na]_o$ increases, along a Michaelis-Menten type curve with a half-maximal rate occurring when $[Na]_o$ equals about 30 mM. The cardioactive steroid-insensitive $[Na]_o$-dependent Na efflux reaches a half-saturation value at $[Na]_o = 38$ mM.[2] Thus two kinds of $[Na]_o$-dependent Na efflux occur in muscle fibers with similar values of $k_{1/2}$, one kind insensitive to and the other kind sensitive to cardioactive steriods.

Above 7°C, the Q_{10} of both strophanthidin-sensitive and strophanthidin-insensitive Na efflux components is about 2.0, whereas below 7°C the strophanthidin-insensitive component has the higher Q_{10}.[35] The two kinds of $[Na]_o$-dependent Na efflux appear to arise from different mechanisms.

Normally the Na influx in muscle fibers is insensitive to ouabain. In a K-free medium, however, Na influx is reduced about 20 percent by ouabain.[19] As both Na influx and efflux are reduced by about the same amount in K-free media, the cardioactive steroid-sensitive Na efflux under these conditions is believed to represent a Na:Na exchange mode of the Na pump. The Na pump in partially poisoned squid giant axons shows a similar behavior.[34,36] Normal red blood cells in K-free media also respond in this way.[37] In red cells the Na:Na exchange is abolished and converted to Na:K exchange by addition of external K. The same behavior is shown by muscle fibers.[19] The

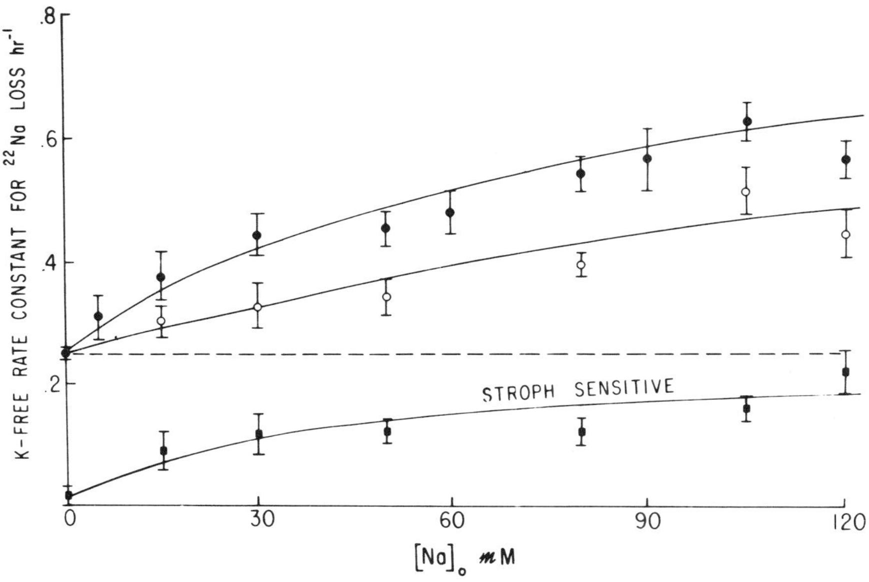

Figure 5.5. The rate constant for loss of ^{22}Na ions from frog sartorius muscle in K-free media is plotted against $[Na]_o$, tris being used to balance the solutions osmotically. The top curve represents the total measured rate constant. The middle curve refers to measurements made in the presence of 10^{-5} *M* strophanthidin. The bottom curve refers to the amount of Na efflux that is abolished by strophanthidin. The dashed line refers to the average value of the rate constant measured in both a K- and Na-free medium. From R. A. Sjodin, *J. Gen. Physiol.* **57,** 164 (1971).

cardioactive steroid-sensitive Na efflux occurring in K-free media in muscle fibers and in the other cell types appears to be due to a one-for-one Na:Na exchange generated by the Na pump. To clarify the terminology, this type of process is referred to as Na pump-exchange diffusion to distinguish it from the kind of Na exchange diffusion first discussed.

In Figure 5.5, a small component of strophanthidin-sensitive Na efflux occurs in solutions that are both K-free and Na-free. An apparently uncoupled cardioactive steroid-sensitive Na efflux of this sort has also been observed in K- and Na-free media when Mg is used to replace Na.[17] A large strophanthidin-sensitive Na efflux was observed in a K- and Na-free medium in which Na was replaced by tris.[38] One cannot conclude that this is evidence for an uncoupled flux by the Na pump, however, since back-pumping of K ions has been shown to occur in K-free media.[1] The different results can be explained by differences in the rate of K leakage from the fibers in K-free media.

THE INFLUENCE OF $[Na]_i$ ON SODIUM EFFLUX

The loss of labeled Na from muscle fibers in a given solution follows single-exponential kinetics, provided that $[Na]_i$ is not permitted to become too low. Thus muscles with an elevated Na content and muscles in a steady state at the normal value of $[Na]_i$ also show an exponential loss of labeled Na after an initial washout period of approximately 20 minutes. In a study of Na exchange diffusion, it was found that prolonged exposure to Li Ringer's solution resulted in a declining rate constant for loss of ^{24}Na and a curve for loss of radioactivity that bends away from the time axis on a semilogarithmic plot.[2] In the Li solution, there is a progressive lowering of $[Na]_i$ due to Na efflux without an accompanying Na influx. If a log-log plot against $[Na]_i$ is made of the loss of ^{24}Na from muscle fibers to Na-free solutions in which Na has been replaced by either Li or choline, a slope of about 3 is obtained.[2,39] The result has led to the "cubic law" for Na efflux from muscle cells according to which an Na pump carrier molecule forms the complex Na_3X before Na transport can occur, where X refers to the carrier. The equation governing Na efflux, as obtained by Mullins and Frumento,[39] is

$$m_o = \frac{k_1}{1 + 1/k_2[\mathrm{Na}]_p^3} \tag{66}$$

where k_1 and k_2 are constants and $[Na]_p$ is the value of $[Na]_i$ at the intake to the Na pump on the inside. The value of $[Na]_p$ differs from bulk $[Na]_i$ owing to the presence of a zone of restricted diffusion, which separates the inner membrane surface containing the Na pump sites from the bulk of the fiber interior. The cubic region of the pump defined by Mullins and Frumento is the region $[Na]_i = 3\text{-}8\ mM$.[39] Above $[Na]_i = 8\ mM$, Na efflux has a linear dependence on $[Na]_i$ as long as $[Na]_i$ is below the region of saturation effects. In the linear region of the Na pump, either a different law is operative or an average of two of the pump sites are already filled and the occupancy of the third site has a linear dependence on $[Na]_i$. Total Na efflux in Na-containing media also consists of an Na:Na exchange diffusion component. This component of Na efflux itself has a linear dependence on $[Na]_i$. The behavior of Na efflux in a Na-free medium where tris is the substitute for Na_i is illustrated in Figure 5.6. The addition of 10 mM K to the solution markedly stimulates the Na pump, which then maintains a linear dependence on [Na]$_i$ down to the cubic region beginning at about 8 mM. Below [Na]$_i$ = 8 mM, the loss of labeled Na becomes much slower with a continuously declining rate constant. As the specific activity of the labeled Na cannot change during the time the muscle is in the Na-free medium, the total Na within the muscle can be calculated assuming that all the muscle Na was initially exchanged with isotope, as is probable. At the four hour point in Figure 5.6, Na ions are

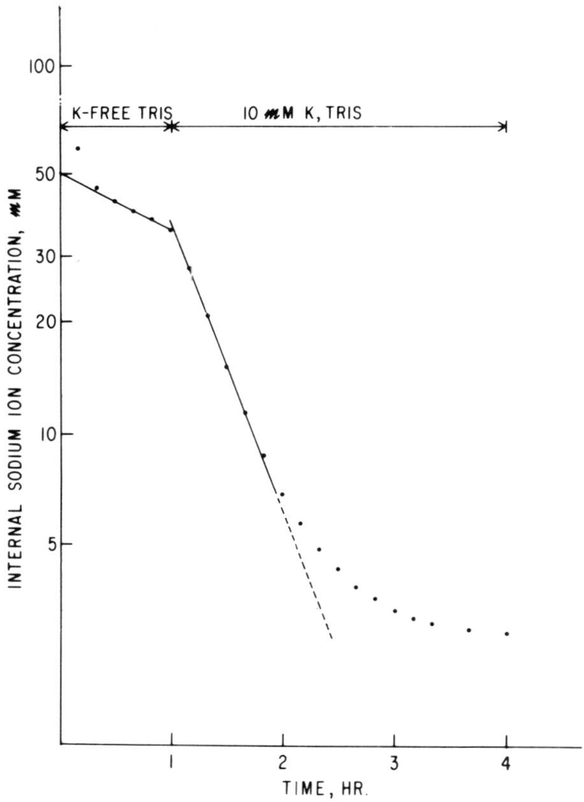

Figure 5.6. Internal Na concentration of frog sartorius fibers is plotted semilogarithmically against time in contact with an Na-free, tris-substituted solution. The muscle was previously fully equilibrated with a Ringer solution with ^{22}Na at known specific activity. Potassium at 10 mM concentration was added to the solution at the time indicated. From R. A. Sjodin, *J. Gen. Physiol.* **57,** 164 (1971).

lost from the muscle at a rate lower than that predicted by the cubic relation. This result would be expected if some fiber Na is located in a different cellular compartment. The residual Na efflux in muscles exposed to Na-free solutions for long periods of time is quite sensitive to the reintroduction of Na to the medium.[2,19] The addition of Na under these conditions causes the residual Na efflux to increase several fold. As this effect is obtained in the presence of ouabain, the results suggest that the ouabain-insensitive exchange diffusion shown by muscle fibers may take place in a compartment other than the sarcoplasm.

ACTIVATING EFFECTS OF FOREIGN CATIONS

Cations not present under normal circumstances are able to activate isolated membrane ($Na^+ + K^+$)-ATPase to different degrees.[40] It is not surprising, therefore, to find that foreign cations are capable of activating the Na pump in cells to varying degrees. In squid giant axons, for example, Rb and Cs ions have a K-like activating effect when present externally, the selectivity order being K, 1:Rb, 0.84:Cs, 0.22.[41] In frog skeletal muscle fibers, both Rb and Cs ions have an activating effect on the Na pump when present externally.[42-44] Rubidium and cesium ions have to be present at higher concentrations than K ions to result in the same degree of activation of the Na pump. The degrees of pump activation produced by $[K]_o = 5$ mM and $[Cs]_o = 25$ mM are roughly equivalent, as shown in Figure 5.7. As is the case for fragmented membrane ($Na^+ + K^+$)-ATPase preparations, NH_4^+ ions also activate the Na pump in skeletal muscle fibers via a K-like effect.[45] Thallium ions activate ($Na^+ + K^+$)-ATPase[45a] and transiently increase Na efflux from frog muscle when $[Tl^+] = 1$ mM.[45b]

As Li ions have been extensively used in past investigations of Na exchange diffusion, the action of Li on the Na pump is of importance. When Li ions are substituted for Na to demonstrate exchange diffusion, the usual effect becomes reversed at high values of $[Na]_i$, unless the Na pump has been inhibited with a cardioactive steroid.[15,19] In addition, the external presence of Li ions in a nominally K-free solution promotes a net Na extrusion against an electrochemical gradient and results in an $[Na]_i$-dependent and ouabain-sensitive Li influx.[15,17] These findings strongly suggest that Li ions have a direct activating effect on the Na pump and are themselves transported into the fibers. However, some caution must be exercised in making this inter-

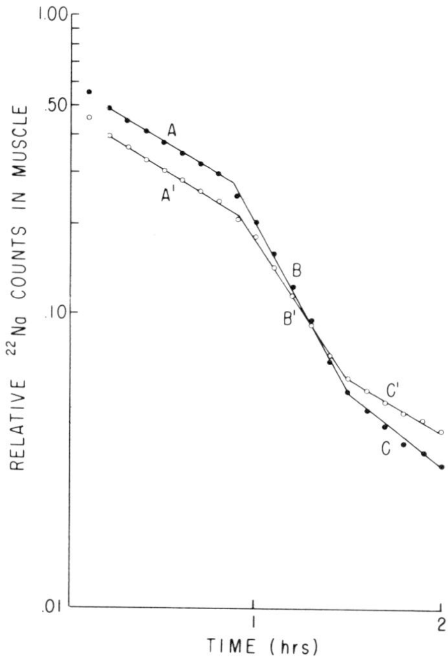

Figure 5.7. Recovery of Na-rich frog sartorius muscles in Ringer's solutions with 5 m*M* K (filled circles, *B*) or 25 m*M* Cs (opened circles, *B'*). Curves *A* and *A'* were obtained in the absence of both K and Cs ions from the bathing medium. Curves *C* and *C'* were obtained in the presence of $10^{-5}M$ strophanthidin. From L. A. Beaugé and R. A. Sjodin, *J. Physiol.* **194,** 105 (1968).

pretation as Li ions have been shown to increase the K efflux in a K-free medium.[33] The additional K leaking from the fibers could clearly result in Na pump activation, especially in an Na-free medium in the absence of an inhibitory effect of $[Na]_o$.[1,16] Some, or even all, of the apparent activating effect of Li on the Na pump in muscle could be due to the effect of K ions leaking from the fibers in view of the fact that such K ions are recaptured by the Na:K pump in Na-free media.[1] The problem was addressed by Beaugé[18] who observed an activating effect of Li on the Na pump even when the K leaking from the fibers was substantially reduced by applying Ba ions externally. Beaugé also measured a Li-activated Na efflux in muscles in which the K content had been reduced to only a few percent of the normal value. Such muscles have a very low K leakage rate, and it seems clear that Li ions do have a direct activating effect on the Na pump in muscle fibers apart from their action in increasing K efflux. As considerable experimental alteration is necessary to demonstrate the direct Li action, it is still difficult to say how much of the increased ouabain-sensitive Na efflux observed in the presence of $[Li]_o$ is due to Li directly and how much is due to K recapture. It seems likely, however, that the direct and indirect effects of Li are roughly on a half and half basis.[18] Though Li ions external to the membrane are K-like in their activation of the Na pump, internal Li ions are unable to exert a Na-like effect, at least to a readily measurable extent.[46] Muscles loaded with Li ions to a high concentration were unable to produce a net Li extrusion into a high-K, Li-substituted recovery solution. Similar results are obtained in other tissues. The external activation sites normally activated by K also respond to Li^+, NH_4^+, Rb^+, and Cs^+, whereas the internal Na site is very specific for Na ions.

THE INDEPENDENCE OF THE SODIUM PUMP IN MUSCLE ON E_m

The Na pump is theoretically capable of both producing an electrical potential and being influenced by a change in the membrane potential. The first possibility is termed "electrogenicity" and is discussed in detail in the following chapter. In fibers isolated from the frog's semitendinosus muscle, the Na pump was not much affected by external K until $[K]_o$ was elevated beyond 7.5 mM to a level that depolarized the membrane considerably.[7] Also, small concentrations of azide that produced membrane depolarization of less than 22 mV were without effect on Na efflux, whereas larger concentrations stimulated the Na efflux.[47] As both K and azide had similar effects on Na efflux and on E_m, it was concluded that membrane depolarization is the immediate activator of the Na pump in muscle and that the activation obtained with K ions is indirect via the membrane potential. In other studies using frog sartorius muscles, however, membrane depolarization alone was unable to activate the Na pump, and changing the value of E_m did not alter the manner in which pump activation depends upon $[K]_o$.[48,49] The muscle fibers were exposed to a high KCl concentration that produced chloride loading and rendered E_m insensitive to changes in $[K]_o$ (see Chapter 4). Under these conditions, the membrane depolarizes to potentials around -22 mV and E_m remains clamped at that value during changes in $[K]_o$, provided that $[Cl]_o$ does not change. The depolarization to -22 mV from the normal value of

E_m failed to produce any pump activation in a K-free medium.[49] Varying $[K]_o$ with E_m clamped produced the same kinetics of pump activation observed when E_m varied with $[K]_o$. The addition of azide produced large depolarizations and failed to activate the pump to any degree in fresh muscles. Azide produced an abnormal, slowly developing pump activation in Na-enriched muscles that was independent of $[K]_o$ but dependent on $[Na]_o$.[49] This kind of pump activation is discussed in the following chapter. The results all indicate that E_m plays no significant role in pump activation in normal sartorius muscles and that activation by $[K]_o$ is independent of E_m. These findings are similar to those in other tissues.[50-52] No explanation is available for the differences in behavior observed in isolated fiber bundles from frog semitendinosus muscle.

SODIUM FLUXES IN OTHER KINDS OF MUSCLE

The description of Na ion movements in skeletal muscle in this chapter follows almost entirely from work on frog muscle. The early accumulation of ion flux data in frog muscle and its widespread use by many investigators have facilitated the development of a general model for Na transport. The fact that the Na pump in frog muscle shows kinetic properties similar to those observed in red blood cells, nerves, and $(Na^+ + K^+)$-ATPase studies places skeletal muscle well within the framework of the classical Na pump. Unfortunately, heavy reliance on a single preparation always raises the question of the generality of results. For this reason, some attention to results using other preparations is necessary.

Sodium fluxes have been investigated in single large invertebrate muscle fibers from the barnacle *Balanus nubilus*[53] and the spider crab *Maia squinado*.[54,55] In both of these preparations, Na efflux is sensitive to external K, internal Na, and external ouabain or strophanthidin. In barnacle muscle fibers, strophanthidin blocks up to 90 percent of the Na efflux, which is considerably more than the amount abolished by removal of external K (about 50 percent).[53] The fact that large reductions in Na efflux occur in this preparation when Na ions are removed from the external medium[53] and that strophanthidin sensitivity is so high suggests that a cardioactive, steroid-sensitive, and Na_o-activated Na efflux occurs in barnacle muscle fibers. This finding is similar to results described in frog muscle, but the details may differ as K was present in the external medium of barnacle fibers when external Na was removed and replaced with either Li or sucrose.[53] In frog muscle, K_o removes the ouabain-sensitive, Na_o-activated Na efflux.

In neither barnacle muscle[53] nor crab muscle[54] do inhibitors of the Na pump reduce Na efflux to the low value expected for passive Na efflux. In view of the large internal electrical negativity, −68 mV in the case of barnacle fibers, passive Na efflux should be no more than about 1 pmole/cm^2·sec. Total Na efflux is around 40 pmole/cm^2·sec so that the largest strophanthidin inhibition observed still leaves around 3 or 4 pmoles/cm^2·sec of residual Na efflux.[53]

Some of the Na efflux may be a strophanthidin-insensitive exchange diffusion, as seen in frog muscle. However, various combinations of K_o and Na_o removal and strophanthidin addition failed to yield lower values of the Na efflux.[53] Indeed, the strophanthidin-insensitive fraction of Na efflux rises to around 40 percent of total efflux of Na when $[Na]_i$ is increased by microinjection. Another observation at elevated $[Na]_i$ is that the rate constant for Na efflux declines indicating a saturation of the Na efflux. This also occurs in crab muscle fibers.[55]

In barnacle fibers, it is the strophanthidin-sensitive fraction that saturates while the strophanthidin-insensitive Na efflux continues a linear rise.[53] In a further study of the effect of inhibitors on Na efflux in muscle (crab fibers), a variety of metabolic inhibitors, which should have inhibited most of the known metabolic pathways for high energy compound formation, failed to reduce Na efflux to calculated passive values.[54] The results led, however, to the suggestion that the Na pump of *Maia* fibers can be divided into two parts, one part driven by ATP via (Na^+ + K^+)-ATPase and the other by respiration and possibly a direct coupling of glycolytic energy.

Additional work is necessary to see if details of Na-pump operation in single-invertebrate muscle fibers are similar to those in frog muscle. As a general rule, however, the basic features of Na-pump operation, whatever the preparation, are clearly the same.

Sodium ion fluxes have also been investigated in mammalian skeletal muscle fibers. Students of human physiology and investigators with medical interests would probably find such studies more interesting than frog muscle studies. However, difficulties with the use of isolated mammalian muscle have been reported[56,57] due to the relatively high rate of oxygen consumption of this tissue, which is some 40 times higher than the rate of O_2 consumption measured in frog muscle. The problem of maintaining mammalian muscle in a steady state during flux measurements is thus much more severe. The results of anoxia on excised mammalian muscle are K loss, Na gain, and membrane depolarization.[56,57] The use of thin diaphragm muscle from the rat with a special O_2 jet directed at the tissue has been reported to overcome some of this difficulty.[57] However, the usual method of bubbling O_2 gas through saline, which then externally perfuses the tissue, is inadequate as mammalian muscle has been reported to gain Na and lose K under such conditions.[56] It thus remains the burden of the investigator to demonstrate adequate oxygenation by careful controls in ion transport studies in mammalian muscle.

Despite all these difficulties, useful results have been obtained with mammalian muscle by paying careful attention to the problem of adequate oxygenation.[58] Using thin diaphragm muscle, researchers have maintained steady internal Na concentration in excised muscle at values around 10 mM. The Na flux values were 10 pmole/cm^2·sec, which compares with a value of around 6 pmole/cm^2·sec for frog muscle at the same value of $[Na]_i$ (Figure 5.1). The value deduced for P_{Na} in rat diaphragm muscle is 2.0–3.6 $\times$ 10^{-8} cm/sec, which is several times higher than the value cited for frog muscle. A criticism of the measurement of P_{Na} in rat muscle is that it is obtained from Na-exchange measurements rather than from net fluxes as is done in frog muscle. Since no measurements are given for Na-free media, an unknown contribution from Na-exchange diffusion exists. The presence of exchange diffusion would of course give spuriously high values of P_{Na} unless appropriate correction were made.

Some data is also available on Na-pump action in mammalian skeletal muscle. In muscle from the rat, ouabain inhibits a component of Na efflux.[59,60] In rat soleus muscle, 10^{-3} M ouabain inhibits 70 percent of the Na efflux,[60] whereas in rat extensor digitorum longus muscle inhibition is less.[59] For rat soleus muscle, the ouabain-sensitive Na extrusion rate was around 0.4 μmole/g·min, which does not differ greatly from Na extrusion rates in frog muscle under similar conditions. The ouabain-sensitive Na extrusion rate was about 1.5 times the ouabain-sensitive ^{42}K uptake rate.[60] This suggests that three Na ions are transported by the pump per two K ions—on the assumption that all the ouabain-sensitive ^{42}K uptake occurs directly via the pump. When more Na ions are extruded than K ions are taken up by the pump, however, the pump contributes to the membrane potential. Such Na pumping is termed "electrogenic pumping." Electrogenic Na pumping can give rise to an ouabain-sensitive K influx via passive K flux channels. If the K conductance is high, such ion movement can be considerable.

In general, ouabain-sensitive K influx cannot be taken as a measure of pumped K influx in tissues showing electrogenic Na pumping unless it can be shown that the electrical contribution is negligible. Nevertheless, it appears that the Na pump in mammalian muscle does not differ greatly from that in frog muscle. However, the distribution of pump sites between SR and surface membranes may vary in different types of muscle. In rat muscle, Na appears to be distributed mainly in two pools; sarcoplasm and the sarcoplasmic reticulum, with Na interchange occurring between the two pools. In this case, much of the Na-pump action appears to be situated in the membranes of the sarcoplasmic reticulum.[59]

REFERENCES

1. R. A. Sjodin and L. A. Beaugé, *J. Gen. Physiol.* **61,** 222 (1973).
2. R. D. Keynes and R. C. Swan, *J. Physiol.* **147,** 591 (1959).
3. E. J. Harris, *Trans. Faraday Soc.* **46,** 872 (1950).
4. R. D. Keynes, *Proc. Roy. Soc. B* **142,** 359 (1954).
5. H. B. Steinbach, *J. Biol. Chem.* **133,** 695 (1940).
6. C. Edwards and E. J. Harris, *J. Physiol.* **135,** 567 (1957).
7. P. Horowicz and C. J. Gerber, *J. Gen. Physiol.* **48,** 489 (1965).
8. H. B. Steinbach, *J. Gen. Physiol.* **44,** 1131 (1961).
9. E. J. Conway and D. Hingerty, *Biochem. J.* **42,** 372 (1948).
10. J. E. Desmedt, *J. Physiol.* **121,** 191 (1953).
11. P. A. Matchett and J. A. Johnson, *Fed. Proc.* **13,** 384 (1954).
12. J. A. Johnson, *Am. J. Physiol.* **187,** 328 (1956).
13. R. A. Sjodin and L. A. Beaugé, *J. Gen. Physiol.* **52,** 389 (1968).
14. R. D. Keynes, *J. Physiol.* **184,** 31P (1966).
15. L. A. Beaugé and R. A. Sjodin, *J. Gen. Physiol.* **52,** 408 (1968).
16. R. A. Sjodin, *J. Gen. Physiol.* **57,** 164 (1971).
17. L. A. Beaugé and O. Ortiz, *J. Physiol.* **226,** 675 (1972).
18. L. Beaugé, *J. Physiol.* **246,** 397 (1975).

19. R. D. Keynes and R. A. Steinhardt, *J. Physiol.* **198,** 581 (1968).
20. A. K. Sen and R. L. Post, *J. Biol. Chem.* **239,** 345 (1964).
21. R. Whittam and M. E. Agar, *Biochem. J.* **97,** 214 (1965).
22. P. J. Garrahan and I. M. Glynn, *J. Physiol.* **192,** 217 (1967).
23. L. J. Mullins and F. J. Brinley, Jr., *J. Gen. Physiol.* **50,** 2333 (1967).
24. M. Dydynska and E. J. Harris, *J. Physiol.* **182,** 92 (1966).
25. J. C. Skou, *Biochim. Biophys. Acta* **23,** 394 (1957).
26. R. L. Post, C. R. Merritt, C. F. Kinsolving, and C. D. Albright, *J. Biol. Chem.* **235,** 1796 (1960).
27. E. T. Dunham and I. M. Glynn, *J. Physiol.* **156,** 274 (1961).
28. L. J. Mullins and F. J. Brinley, Jr., *J. Gen. Physiol.* **53,** 704 (1969).
29. J. D. Robinson, *Arch. Biochem. Biophys.* **139,** 17 (1970).
30. G. E. Lindenmayer, A. Schwartz, and H. K. Thompson, *J. Physiol.* **236,** 1 (1974).
31. R. L. Post, S. Kume, T. Tobin, B. Orcutt, and A. K. Sen, *J. Gen. Physiol.* **54,** 306S (1969).
32. G. J. Siegel and R. W. Albers, Nucleoside Triphosphate Phosphohydrolases, in *Handbook of Neurochemistry,* Vol. 4, Plenum Press, New York (1970), p. 13.
33. G. McClellan, H. Gonzalez-Serratos, and P. Horowicz, *Fed. Proc.* **33,** 1458a (1974).
34. P. F. Baker, M. P. Blaustein, R. D. Keynes, J. Manil, T. I. Shaw, and R. A. Steinhardt, *J. Physiol.* **200,** 459 (1969).
35. P. Horowicz, J. W. Taylor, and D. M. Waggoner, *J. Gen. Physiol.* **55,** 401 (1970).
36. P. C. Caldwell, A. L. Hodgkin, R. D. Keynes, and T. I. Shaw, *J. Physiol.* **152,** 591 (1960).
37. P. J. Garrahan and I. M. Glynn, *J. Physiol.* **192,** 159 (1967).
38. E. G. Henderson, *Life Sci.* **10,** Part I, 767 (1971).
39. L. J. Mullins and A. S. Frumento, *J. Gen. Physiol.* **46,** 629 (1963).
40. J. C. Skou, *Biochim. Biophys. Acta* **42,** 6 (1960).
41. R. A. Sjodin and L. A. Beaugé, *J. Gen. Physiol.* **51,** Part 2, 152s (1968).
42. R. A. Sjodin, *J. Gen. Physiol.* **44,** 929 (1961).
43. R. H. Adrian and C. L. Slayman, *J. Physiol.* **184,** 970 (1966).
44. L. A. Beaugé and R. A. Sjodin, *J. Physiol.* **194,** 105 (1968).
45. L. A. Beaugé and O. Ortiz, *J. Exp. Zool.* **174,** 309 (1970).

45a. C. Hegyvary and R. L. Post, *J. Biol. Chem.* **246,** 5234 (1971).

45b. L. J. Mullins and R. D. Moore, *J. Gen. Physiol.* **43,** 759 (1960).

46. R. D. Keynes and R. C. Swan, *J. Physiol.* **147,** 626 (1959).

47. P. Horowicz and C. J. Gerber, *J. Gen. Physiol.* **48,** 515 (1965).

48. L. A. Beaugé, R. A. Sjodin, and O. Ortiz, *Biochim. Biophys. Acta* **389,** 189 (1975).

49. L. A. Beaugé and R. A. Sjodin, *J. Physiol.* **263,** 383 (1976).

50. F. J. Brinley and L. J. Mullins, *Ann. N.Y. Acad. Sci.* **242,** 406 (1974).

51. D. Cotterrell and R. Whittam, *J. Physiol.* **214,** 509 (1971).

52. M. F. Marmor, *J. Physiol.* **218,** 599 (1971).

53. F. J. Brinley, Jr., *J. Gen. Physiol.* **51,** 445 (1968).

54. E. E. Bittar, *J. Physiol.* **187,** 81 (1966).

55. E. E. Bittar, P. C. Caldwell, and A. G. Lowe, *J. Marine Biol. Assoc. U.K.*, **47,** 709 (1967).

56. R. Creese, *Proc. Roy. Soc. B* **142,** 497 (1954).

57. R. Creese, N. W. Scholes, and W. J. Whalen, *J. Physiol,* **140,** 301 (1958).

58. R. Creese, *J. Physiol.* **197,** 255 (1968).

59. E. Rogus and K. L. Zierler, *J. Physiol.* **233,** 227 (1973).

60. T. Clausen and O. Hansen, *Biochimica et Biophysica Acta* **345,** 387 (1974).

6

Electrogenicity and Modes of Coupling of the Sodium-Potassium Pump in Muscle

When the value of E_m is less negative than the value of E_K, an inwardly directed K pump is required to maintain a steady state for K ions. This situation applies to normal freshly dissected muscle fibers. On the other hand, aged muscle fibers that have lost K ions in exchange for Na ions may regain the lost K ions. This can occur by two different mechanisms. First, an increased rate of inward K pumping can, by exceeding the K efflux, produce a net influx to restore the fiber's K level. Second, if the Na and K pumps are coupled such that more Na ions are extruded per pump cycle than K ions are accumulated, the pump can produce a hyperpolarization whereby E_m now becomes more negative than E_K. The resulting passive net influx of K restores the K level. In addition, both of these mechanisms can operate simultaneously.

ELECTROGENICITY OF THE SODIUM PUMP IN MUSCLE FIBERS

The mechanism by which the Na:K pump produces an electrical potential is called electrogenic pumping. The ability of the Na pump in muscle fibers to develop an electrical potential was first clearly demonstrated by Kernan,[1] who showed that the value of E_m was more negative than E_K in frog sartorius fibers recovering from the Na-rich state in the presence of 10 m*M* of external K. The increased negativity within the fibers correlated well with the extrusion of Na ions because, at the end of the period of Na extrusion, the values of E_m and E_K agreed to within experimental errors. Kernan also showed that Cl ions did not contribute to the increased negativity within the fibers, as similar results were obtained in a Cl-free Ringer's solution in which Cl ions were replaced by SO_4 ions.

In a parallel study, it was shown that the electrical potential developed by the Na pump in muscle fibers is temperature-sensitive and increases with the magnitude of the net Na outward flux produced by the fibers.[2] The measured change in potential created by the Na pump, corrected for the contribution of the term (RT/F) to the potential, was around -40 mV at the highest Na extrusion rates observed. The electrical negativity generated by the pump declined with decreasing Na pump flux along a curve that extrapolated to no change in potential at zero net Na pump flux. The potential produced by the pump developed rapidly when fibers were warmed from 3° to 25°C, and reached a maximum value before beginning to decline. In addition, Frumento showed that the maximum change in potential occurred in a time so short that internal ion concentrations did not have sufficient time to change significantly.[2] Similar results were obtained by Mullins and Awad,[3] who found that muscle fibers with normal values of $[Na]_i$ showed only a small increase in E_m upon warming from 4° to 20°C, whereas E_m for Na-enriched muscles became more negative by about 30 mV upon warming to 20°C. E_m in fibers transferred to a solution containing 2.5 m*M* [K] was found to be initially 13 mV more negative than E_K. Thus it was concluded that the Na pump of skeletal muscle fibers is capable of generating an electrical potential.

Additional evidence that the temperature-dependent hyperpolarization observed in Na-enriched sartorius muscle fibers is associated with Na pump activity is provided by the fact that ouabain abolishes the hyperpolarization.[4-6] As ouabain also abolishes net Na extrusion from Na-enriched muscles, the correlation between the net Na flux produced by the pump and the additional negativity within the fibers is again good. Taken together, all of the evidence points to the existence of an electrogenic Na pump in the muscle fiber membrane that can, under appropriate conditions, develop additional electrical negativity within the fibers of up to around −40 mV. As the electrogenic potential depends upon the rate of Na extrusion, it also has a strong dependence on the value of $[Na]_i$. Due to the high membrane conductance to K ions, whenever the Na pump produces extra negativity within the fibers, there will exist an extra influx of K via passive flux channels whether or not a K pump is operative. As the electrogenic hyperpolarization is ouabain sensitive, the associated extra influx of K through passive channels will also be abolished by ouabain.

CARDIOACTIVE STEROID-SENSITIVE POTASSIUM FLUXES

Harris' observation[7] of a marked slowing effect of ouabain on K influx into frog sartorius muscle indicates that significant pump-related K uptake occurs under certain conditions. This shows that, in addition to the kind of K influx discussed in Chapter 4, which is relatively insensitive to ouabain, additional components of pump-related K influx can occur. To avoid confusion, the terminology applied to K influx must be clarified and carefully defined. It is important to distinguish the various possibilities. A K pump may be present; it may or may not be coupled to the Na pump. The Na pump may create an electrogenic potential that adds additional negativity to the membrane potential. This additional negativity can lead to an increase in passive K influx. Pump-related K influx, therefore, may be due to pumped components of K flux or simply to an increase in the K influx through already existing channels if an electrogenic component of E_m is present. Both sources of pump-related K ion movement may, of course, be present simultaneously. It should be realized that ouabain alone cannot be used to distinguish between these different modes of pump-related K flux. The amount of K influx inhibited by ouabain only gives the total pump-related K influx.

In freshly dissected muscle, it was found that ouabain had no significant effect on K influx.[8] The reason for the variation between the two sets of results becomes apparent when the influence of $[Na]_i$ is taken into account.[9] By storing freshly dissected frog sartorius muscles in 105 m*M* Na, 5 m*M* K Ringer's solution for a three hour period at 20°C, a uniform, low Na concentration of about one-half the normal value is obtained.[9] Muscles with varying elevated Na concentrations are easily prepared by storing them in K-free Ringer's solution at 4°C for different lengths of time. The magnitude of K influx is over 60 percent higher in the high $[Na]_i$ muscles than in the low $[Na]_i$ muscles. In the muscles with low $[Na]_i$, K uptake was marginally insensitive to the external application of 10^{-5} M strophanthidin, the average reduction in K influx being about 5 percent. In the muscles with elevated $[Na]_i$, however, strophanthidin reduced K influx by 45 percent. As strophanthidin reduced K influx in high $[Na]_i$ muscles to about the level observed in the low $[Na]_i$ muscles, all of the extra K influx generated by elevating the value of $[Na]_i$ is strophanthidin sensitive.

The kinetic details of the cardioactive steroid-sensitive K influx are considered in the next section. At this point, it is only noted that this component of K influx has three features in common with the Na pumped efflux; it depends upon $[Na]_i$, it is inhibited by ouabain or strophanthidin, and it varies nonlinearly with $[K]_o$ in the low concentration range in Na Ringer's solution. It will be recalled that the ouabain- or strophanthidin-insensitive K influx varies linearly with $[K]_o$ at low values of concentration. The strophanthidin-sensitive K influx, however, is 12 times greater when $[K]_o = 5$ mM than when $[K]_o = 1$ mM.

The effect of strophanthidin on K efflux has also been studied.[9] In muscles with low $[Na]_i$, strophanthidin was without effect on K efflux. In muscles with elevated $[Na]_i$, however, strophanthidin increased K efflux by about 70 percent. The strophanthidin sensitivity of K efflux under these conditions was greatly reduced if K ions were removed from the medium. The effect of strophanthidin on K efflux is thus maximal under conditions favoring a high Na pumping rate in the absence of strophanthidin. This component of K efflux appears to be definitely pump related, yet its mechanism remains obscure. An attractive hypothesis is that some K ions leaving the fibers are caught by either the pump itself or by an electrogenic effect due to the pump operation and are subsequently transported back into the fibers. Unfortunately, there are a number of factors that argue against this captivating interpretation. First, the effect is observed in Na Ringer's solution where the inhibitory action of $[Na]_o$ would render ineffective an additional pump activation due to a small K concentration buildup from K ions leaving the cell. Secondly, the effect is strongly present when $[K]_o = 5$ mM and small concentration changes just at the fiber surface should be proportionally small. A third factor is that, in the absence of strophanthidin, the net K flux is in the inward direction, which, if anything, should deplete the K level immediately at the fiber surface. Finally, were this the correct explanation, the magnitude of K efflux in the absence of pump inhibitor should be significantly lowered in muscles with elevated $[Na]_i$. The rate constant for ^{42}K loss under the conditions of these experiments was observed to be independent of the value of $[Na]_i$.[9] As strophanthidin abolishes electrogenic hyperpolarization produced by the pump, the ensuing depolarization could account for the increased K efflux. This explanation could account for the influence of both $[Na]_i$ and $[K]_o$ on the results.

A precise analysis of these results would require a knowledge of the membrane conductance to K ions as influenced by $[Na]_i$, the Na pumping rate, and the addition of pump inhibitor. Unfortunately, conductance data in absolute units is unavailable for these cases. It is of interest, however, that fiber membrane conductance increases in muscles with elevated $[Na]_i$, upon the addition of ouabain.[10] As $[Na]_i$ becomes lower, the ability of ouabain to increase the conductance decreases. If these conductance changes were due to K channels, the effects of pump inhibitor on ^{42}K efflux would find an explanation.

In Na- and K-free media, ouabain or strophanthidin increases K efflux severalfold in muscles with elevated $[Na]_i$.[11,12] In this case, however, the effect can be readily explained in terms of a reaccumulation of K ions leaking out of the fibers in the K-free solution. In Na-free media, the lack of inhibitory action of $[Na]_o$ on the pump enables a small concentration buildup of K just outside the fiber membrane to activate the pump and bring about the reuptake of K ions leaving the fiber. This ouabain-sensitive K efflux differs from that previously discussed for media containing normal amounts of Na in two respects. First, the rate of K loss in the absence of inhibitor falls to abnormally low values, indicating a slowing effect due to the pump. Second, when ouabain is added the rate constant for ^{42}K loss increases to normal values, resulting in an up to fourteenfold increase in K efflux.[12] Recall that, in Na Ringer's solution, the effect is not nearly as pronounced, and amounts to only a 24 percent increase in K efflux in a K-free solution.

The mechanism for the pump-related reuptake of K can involve electrogenic action of the Na pump, actual reuptake of K ions by a K pump, or both. Which effect predominates seems to depend upon the Na substitute used. When tris is used to replace Na, the pump becomes sensitive to small concentrations of K and ouabain-sensitive hyperpolarizations are observed.[11,13] In the experiments of Henderson, addition of strophanthidin caused a 15 mV depolarization in muscles with elevated $[Na]_i$ in tris Ringer's solution, while

K efflux was increased approximately fourteenfold.[11] A combination of depolarization and K conductance increase could conceivably account for the results. In the absence of accompanying membrane conductance data, no further conclusion can be drawn. When Mg ions are used to replace external Na, however, no significant depolarization occurred upon addition of ouabain, while K efflux still increased fourteenfold.[12] The absence of a significant electrogenic effect in K-free Mg Ringer's solution strongly suggests that K reuptake under these conditions is due solely to a chemically coupled K pump. The latter results may, therefore, be taken as an experimental demonstration of the occurrence of "back-pumping" of K ions in frog muscle fibers.

COUPLING OF THE SODIUM-POTASSIUM PUMP IN MUSCLE FIBERS

From the previous discussion, it is clear that two classes of pump-related K influx can occur in muscle fibers: one, an increased K uptake due to electrogenic action of the Na pump, and two, K entry via a chemically coupled pump. As an electrogenic action of the Na pump has been demonstrated experimentally in muscles recovering from the Na-rich state, such muscles will always show some electrically coupled K uptake due to the finite membrane conductance of K ions. The question is, how much actual K pumping normally occurs in various physiological states of the fibers and can this be determined experimentally with any precision? Clearly, accurate measurements of the membrane conductance to K ions in absolute units for all desired conditions together with a knowledge of the electrogenic hyperpolarizations would permit the isolation of the electrically coupled part of the K influx. From a knowledge of the total K uptake during recovery, the actual K flux produced by a K or Na-K pump could then be determined. Unfortunately, it has not yet been possible to measure the membrane conductance to K ions during Na extrusion with sufficient accuracy to permit application of this method with any real degree of precision. Existing conductance measurements permit one to state only that the electrically coupled K uptake during the recovery of Na-enriched muscles is from half to all of the inward K movement.[4,5]

The crucial factor that determines whether electrically coupled K uptake suffices to explain the results or whether a K pump is required is the value of

E_m with respect to the value of E_K. The membrane potential of Na-enriched fibers at 4°C in the K-free Na loading solution is 90.9 ± 2.1 mV.[6] When the fibers are warmed to 20°C, the membrane potential changes to about −110 mV.[4,6,14] The behavior of E_m upon adding K to the medium depends upon the concentration. When $[K]_o$ is rapidly changed from nominally zero to 10 m*M*, E_m quickly changes to about −70 mV where it remains for the ensuing 45 minutes.[4] The initial value of E_K in the 10 m*M* K recovery solution is about −50 mV. The value of E_K becomes more negative during recovery and E_m and E_K approach one another after about one hour at the end of recovery. During the recovery period, E_m is more negative than E_K and electrical coupling of the Na and K net fluxes suffices to explain the results. When $[K]_o$ is rapidly changed to 5 m*M* from zero to initiate recovery, the value of E_m becomes less negative but at a slower rate. The value of E_m remains at about −85 mV during a large part of this recovery period.[14] In this case, the initial value of E_K is −70 mV, again less negative than E_m, and E_m and E_K again approach each other during recovery.

In the cases just discussed, a significant part of the reaccumulation of K during recovery can be attributed to electrical coupling. As a matter of fact, almost all of the net K uptake could be due to electrogenic action of the Na pump. In certain other cases of fibers recovering from Na enrichment, this is not the case. If the muscle fibers are loaded with Cl as well as Na prior to recovery, it is found that K reaccumulation now occurs during a period in which E_m is less negative than E_K.[15] Under these conditions, the passive net K flux is an outward leakage; in this case, there can be no electrogenic contribution of the pump to net K uptake. One must regard the K reaccumulation as being due to a chemically mediated K pump. In these experiments, Cl loading can be effected by exposing the Na-loaded fibers to 10 or 30 m*M* KCl Ringer's solution at 1°C for an hour prior to the recovery period at 20°C. The low temperature prevents significant recovery from taking place. Due to the ensuing Cl loading, the value of E_m is held at −30 to −40 mV during the initial phase of recovery, at which time E_K is approximately −50 mV in 10 m*M* K Ringer's solution.

In this study of Keynes and Rybova, some muscle fibers had low membrane potentials after the period of Na loading even without an initial period of Cl loading.[15] Such muscles also extruded Na and reaccumulated K with E_m being less negative than E_K. The muscles in this category must be regarded as abnormal with respect to their values of E_m after Na loading, as most fibers have high membrane potentials both in the cold in the Na-loading solution, and at 20°C in the K-free Ringer's solution prior to beginning the recovery experiment.[6] Fibers are usually more negative than −100 mV but they become less negative when K is added to the solution, though still remaining more negative than E_K during recovery. In a follow-up to this study, Cross et al. were unable to obtain Na-loaded fibers with low membrane potentials even after Cl loading.[4] They subsequently found that fibers with low membrane potentials could best be achieved by lowering the Ca and Mg concentrations in the solutions or by omitting Ca and Mg altogether. Such muscle fibers underwent a normal recovery with E_m remaining less negative than E_K. These results were grouped according to the value of E_m after 24 hours in the Na-loading solution just prior to beginning the recovery experiment. When E_m was more negative than −50 mV, a normal electrogenic hyperpolarization occurred during recovery, whereas muscles in which E_m was less negative than −50 mV failed to show the hyperpolarization. All muscles, irrespective of the value of E_m, underwent a normal recovery process.

The reason why some muscle fibers show low membrane potentials after Na loading, apart from experimental manipulations, remains obscure. The fact that some do, however, is a fortunate finding, as these muscle fibers provide clear-cut evidence that net K reaccumulation occurs even when E_m is less negative than E_K. This has to be regarded as firm verification of the existence of a metabolically mediated K pump in muscle fibers.

The findings discussed in this section can best be summarized by stating that both electrical and chemical coupling of inward K movement to the Na pump occur in muscle fibers. In the cases where both modes of coupling occur simultaneously, measurements of recovery fluxes, conductance, and E_m have thus far failed to permit a quantitative statement to be made about how much of each coupling mode is present.

COUPLING RATIO OF THE SODIUM-POTASSIUM PUMP DURING RECOVERY AS OBTAINED FROM PUMP: LEAK RATIOS

When the Na pump contributes to the membrane potential, equations can be derived that include the coupling ratio, r, that is, the number of Na ions pumped per K ion pumped.[16] This approach was outlined for steady-state conditions in Chapter 3 (equations 48 to 53). Ionic concentrations, the ratio $\alpha = P_{Na}/P_K$, and E_m can then be used to calculate the coupling ratio. If the value of r is known during recovery, the previous question of how much K reaccumulation is electrically coupled and how much is chemically coupled to the pump can be answered decisively. Equation (53) is not valid for recovery experiments, however, as it applies only to steady-state conditions. Nevertheless, it can be suitably modified for the non-steady-state conditions obtaining during recovery.[6] Using the same notation as in Chapter 3, one may write

$$f_{Na} j_{Na} = -m_{Na} \tag{67}$$

$$f_K j_K = -m_K \tag{68}$$

where the factors f are ratios of pumped flux to leak flux for each cation. In the case of Na ions, f_{Na} is always positive, as pumped flux and leak occur in opposite directions. The value of f_K may be either positive or negative, however, depending on whether E_m is less negative or more negative than E_K. If the pump generates an electrical potential that makes E_m more negative than E_K, both passive K leak and K pumped flux are in the inward direction, and f_K is negative. Following the same procedure applied in Chapter 3 (equations 48 through 51),

$$j_{Na} = -Y j_K \tag{69}$$

where $Y = (f_K/f_{Na})\,(r)$. Substituting values of j_{Na} and j_K using equation (9) and solving for the membrane potential gives

$$E_m = \frac{RT}{F} \ln \frac{YP_K[\mathrm{K}]_o + P_{Na}[\mathrm{Na}]_o}{YP_K[\mathrm{K}]_i + P_{Na}[\mathrm{Na}]_i} \tag{70}$$

Recalling that $\alpha = P_{Na}/P_K$ and defining the parameter $\theta = \alpha/Y$, equation (70) can be written in the following form:

$$E_m = \frac{RT}{F} \ln \frac{[\mathrm{K}]_o + \theta[\mathrm{Na}]_o}{[\mathrm{K}]_i + \theta[\mathrm{Na}]_i} \tag{71}$$

As f_{Na} and r are always positive quantities, a negative value of f_K results in a negative Y and, hence, a negative value for θ. It is clear that, for these conditions, equation (71) predicts a value of E_m that is more negative than E_K. Equation (71) is thus a general expression that takes into account both permeability to ions and the electrogenic contribution of the Na-K pump for both steady- and non-steady-state conditions. In the steady state, $f_{Na} = f_K = 1$ and $Y = r$, in which case equation (71) becomes identical to equation (53).

Two other useful relations are obtained from the fact that, during recovery, the total net fluxes of Na and K balance each other so that net Cl fluxes are negligible.

$$j_{Na} + m_{Na} = -(j_K + m_K) \tag{72}$$

Substitution of previous relations into equation (72) gives

$$Y = \frac{1 - f_K}{1 - f_{Na}} \tag{73}$$

and

$$r = \frac{f_{Na}(1 - f_K)}{f_K(1 - f_{Na})} \tag{74}$$

for $f_{Na} \neq 1$ and $f_K \neq 1$. Since ion concentrations during recovery are already known, all that is needed for the calculation of θ is a measurement of E_m during recovery. Provided that α is known, Y can also be determined. The factor f_{Na} is known from flux measurements, hence f_K and r can be calculated from equations (73) and (74). The main obstacle in applying this method to the recovery of Na-enriched muscle fibers is a lack of precise knowledge as to the state of P_{Na} and P_K during a period of rapid Na extrusion. For potentials between -70 and -125 mV, P_{Na} at rest is independent of $[Na]_i$ and relatively independent of E_m.[12] Though P_{Na} has not actually been measured during a period of rapid Na extrusion, it appears safe to suppose that it is independent of the Na pump. In the case of P_K, however, the situation is not so clear because the value of P_K often depends upon the value of $E_m - E_K$. The most extensive measurements have been made by Hodgkin and Horowicz, who found P_K to increase with inward driving forces.[17] It is not possible, however, to find measurements of P_K or G_K for conditions that match those holding during the recovery process. The total membrane resistance measured in the presence of ouabain in sulfate Ringer's solution does not change significantly during the Na-loading procedure.[5] This suggests that P_K, too, remains unchanged during recovery, and if this is so, the value of α measured in ouabain can be used in applications of equations (70) or (71). Using the value of E_m obtained with ouabain,[18] α is calculated to equal 0.016. The inward net leakage rate for Na ions is known accurately and has been shown to be independent of $[Na]_i$.[12] The factor f_{Na} is thus known, from kinetic measurements of Na pumping rate, at each point during recovery. This method has been used to calculate the coupling ratio in the beginning phase of recovery from the initial value of E_m during recovery.[6] Via equation (70) or (71), E_m provides a value for Y using the same value for α stated previously. The value of f_K is then calculated from equation (73), permitting r to be calculated from equation (74). The value calculated for r by this means is about 1.5; about 75 percent of the net K uptake is pumped, and about 25 percent is electrically coupled. These assumptions are reasonable and there is experimental justification for each of them. It is evident, however, that if P_K increases during Na pumping, the value of r increases and a greater fraction of the net K movement is electrically coupled.

COUPLING RATIO OF THE SODIUM-POTASSIUM PUMP IN THE STEADY STATE

The advantages of considering muscles in or near the steady state are that there are fewer assumptions to make, there is a closer correspondence to normal physiological conditions, and the value of α is known with greater precision. By considering muscle fibers equilibrated in Ringer's solution for a few hours to be in, or very close to, a steady state, some evidence from potential measurements indicated that $r \geq 3$.[16] The difficulties in precise reckonings of r were pointed out in this work. To obtain these results, the value of α was taken to be 0.01 and the activity coefficient for K inside the fibers was taken to be 0.60, that of K creatine phosphate. However, direct measurements of the activity coefficient for K ions inside the fibers, using K-sensitive glass microelectrodes, give a value of 0.77, which is close to the value in the external medium.[19] No great error results from omitting activity coefficients in equations (39), (53), and similar equations.

It has also been proposed that the coupling ratio of the Na-K pump in muscle fibers may vary under different experimental conditions.[3,4] This point is considered further in the next section.

An approach mentioned in Chapter 3 is to deduce α from net flux data for Na and K ions in the absence of significant cation pumping or from net fluxes corrected for active ion pumping. The value of α from such calculations is 0.02, giving a value for $r = 2.0$ from equation (53) for fresh, non-Na-loaded muscle fibers. Another approach is to use the value of E_m measured in ouabain or strophanthidin to obtain α from equation (39), and the value of E_m in the absence of inhibitors to obtain r from equation (53). For the same value of $[K]_o = 2.5$ mM, E_m in strophanthidin is -86.9 mV,[18] giving a value of $\alpha = 0.016$. For normal fresh muscle fibers with $E_m = -92$ mV (see Table 3.1), the value calculated for r is 1.6. This calculation should not be subject to the criticism that α may have changed, as P_{Na} and P_K should remain the same for the conditions used. This value of r may be expected to apply to muscle fibers that are either in a steady state or are experiencing only slight Na gains and losses. The various values obtained for r are summarized in Table 6.1.

THE KINETICS OF THE POTASSIUM PUMP IN MUSCLE FIBERS STUDIED IN SOLUTIONS CONTAINING BARIUM IONS

Owing to difficulties in separating chemically coupled K pumping from electrically coupled K fluxes, it has not been possible to study the kinetics of the K pump in muscle fibers under normal recovery conditions. If K flux channels could be effectively blocked, the electrically coupled K flux during recovery might be substantially reduced, thus enabling K pumped fluxes to be measured with better resolution. Some of the results presented in Chapter 4 suggest that Ba ions would be a good choice as agents to block K channels. Provided that Ba ions do not interfere with the Na-K pump, the K ion movements during recovery of muscle fibers in solutions containing Ba should be largely due to the K pump. A combination of Ba ions and ouabain should thus block most of the K influx (Figure 6.1). This technique was ap-

Table 6.1. Summary of Values for the Na:K Coupling Ratio

Method	R
From E_m and equation (73) (Na-enriched muscles)	1.5
E_m, steady state (Mullins and Noda)[16]	≥3
α from net flux, α/r from E_m	2.0
$E_m \pm$ ouabain, steady state	1.6
Presence of Ba ions (Na-enriched muscles)	1.1

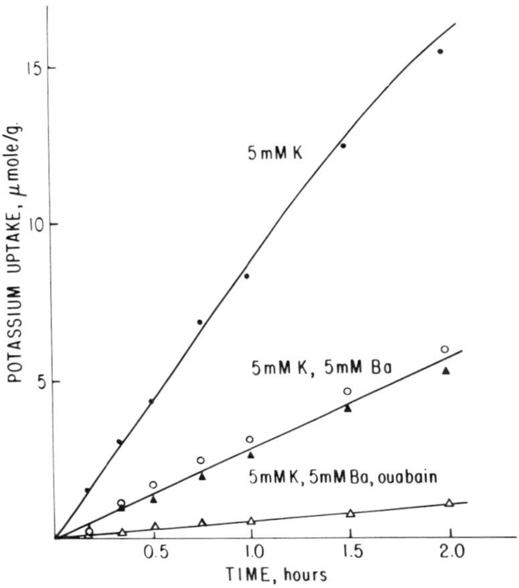

Figure 6.1. Uptake of K by frog sartorius muscles as influenced by Ba and ouabain using ^{42}K. Filled symbols refer to a pair of muscles from the same frog, open symbols to another pair. From R. A. Sjodin and S. C. Wu, unpublished results.

plied by Sjodin and Ortiz, who showed that Na-enriched sartorius muscle fibers underwent a normal recovery in the presence of millimolar concentrations of Ba ions.[6] As the net K influx under these conditions reflects mainly K pump activity, the K influx should be largely ouabain-sensitive. With $[K]_o = 5$ mM, almost 100 percent of the K influx was abolished by ouabain, thus bearing out this prediction (Figure 6.2). It will be recalled that only about 50 percent of the K influx is abolished by ouabain during normal recovery in the absence of Ba.

Calculations based on the values of P_K and G_K in the presence of Ba ions show that over 90 percent of the net K influx is due to K pumping and at most 10 percent is electrically coupled under these conditions. The kinetics of the K pump can thus be effectively measured. The K pumped flux rises with increasing values of $[K]_o$ along an approximately rectangular hyperbolic curve in 105 mM Na Ringer's solution, an S-shape being noted near the origin. When all external Na is replaced by tris, the K pumped flux rises and any apparent sigmoidicity in the kinetics is lost (Figure 6.2). In the absence

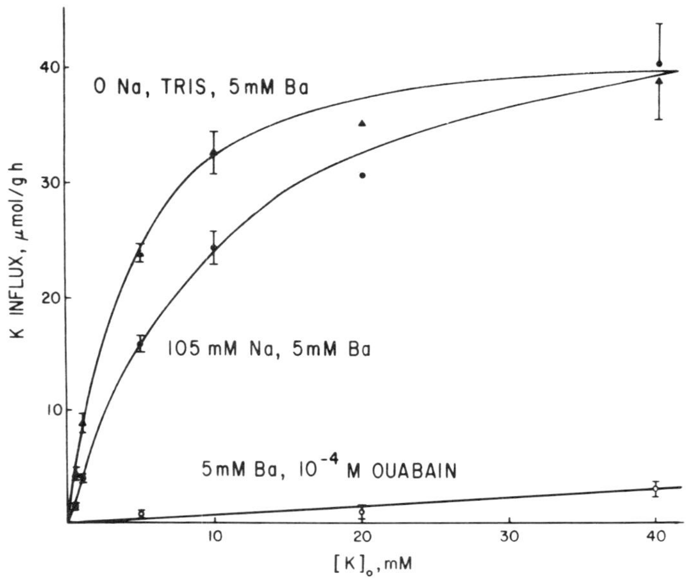

Figure 6.2. Influx of ^{42}K in frog sartorius muscle in the presence of Ba ions is plotted versus $[K]_o$. All muscles were enriched with Na, prior to measurement of influx, by storage in K-free Ringer solution at 4°C for 24 hours. Triangles; Na-free tris substituted Ringer solution; solid circles; Na Ringer's solution; open circles; presence of Ba ions and ouabain. From R. A. Sjodin and O. Ortiz, *J. Gen. Physiol.* **66,** 269 (1975).

of external Na, the data are fit by a rectangular hyperbola, the K pump rate being given by

$$\frac{[\mathrm{K}]_o}{m_{\mathrm{K}}} = \frac{1}{V_{\max}}[\mathrm{K}]_o + \frac{k_m}{V_{\max}} \tag{75}$$

where m_{K} is the K pumped flux, $V_{\max}$ is the maximal K pump flux and k_m is a constant equal to 3.5 m*M*. The K pump data are plotted reciprocally in Figure 6.3 in the form of a modified Lineweaver-Burk plot. As the K pump rate is also a function of Na_i, it has to be stated that these K pump rates apply to fibers in which the average Na content is 30 μmoles/g initially ($[\mathrm{Na}]_i =$ 48 m*M*). As the value of Na_i changes rapidly during recovery, one may speak only of the initial flux values observed when muscles recover from Na loading.

Muscles were loaded to various initial values of $[\mathrm{Na}]_i$ by varying the time

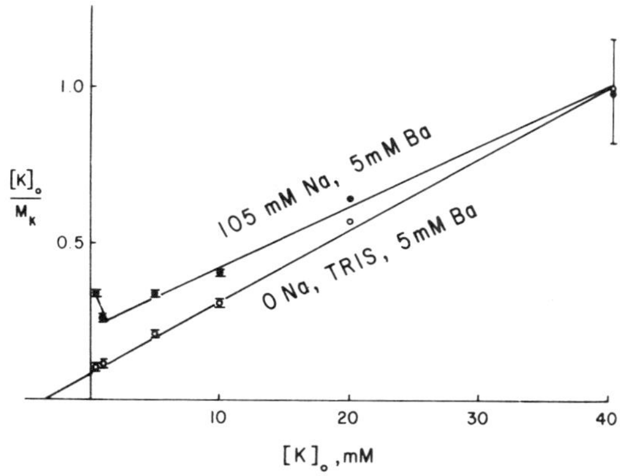

Figure 6.3. Modified Lineweaver-Burk plot of K influx in the presence of Ba ions in both Na and Na-free Ringer's solution for Na-enriched frog sartorius muscle. Potassium influx, M_K, is expressed in units of μmole/g·hr. From R. A. Sjodin and O. Ortiz, *J. Gen. Physiol.* **66,** 269 (1975).

in cold K-free Ringer's solution prior to the recovery experiment. The dependence of the K pump on $[Na]_i$ could then be worked out using Ba ions. The various components of K influx involved are plotted in Figure 6.4. The value of K influx measured in the presence of 5 m*M* Ba may be taken as the K pump rate for the reasons already discussed. Essentially all of this flux is ouabain sensitive, as indicated by the dotted line showing influx in the presence of both ouabain and Ba ions. The figure shows that the K pump rate increases linearly with increasing values of $[Na]_i$.

The findings concerning the kinetics of the K pump in muscle fibers have certain implications with respect to the coupling between the K and Na pumps in muscle. A rigidly coupled Na-K pump requires that the Na and K pumped fluxes have similar kinetics. When the concentration of an activating or modifying cation is changed, for example, rigid Na to K coupling requires that pumped Na efflux and pumped K influx be altered by the amounts calculated from the appropriate value of r. The behavior of pumped K influx when $[K]_o$ is varied (Figures 6.2 and 6.3) is qualitatively similar to the behavior of Na efflux under similar conditions (Figures 5.3 and 5.4). Both sets of curves tend to reach saturation at high values of $[K]_o$ and the values of $k_{1/2}$ are comparable.

In Na-free media, single site saturation kinetics are obeyed for both Na and K active fluxes. The addition of external Na is inhibitory to both the Na

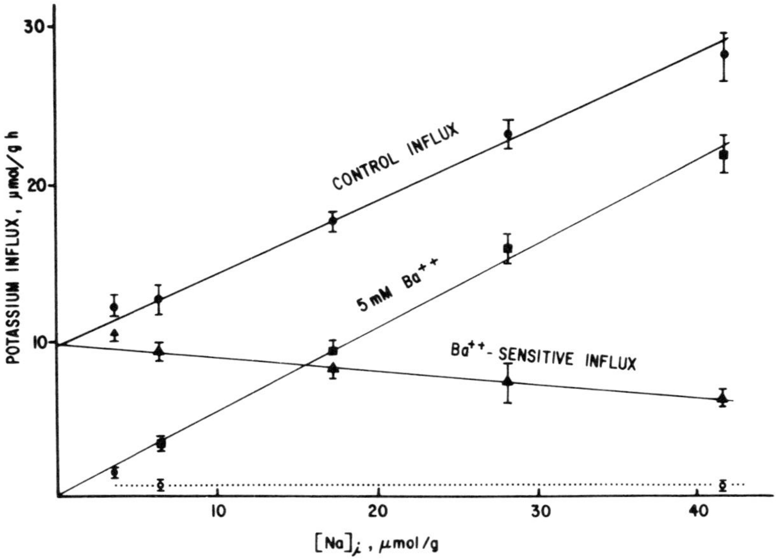

Figure 6.4. Potassium influx and its components in Na-enriched frog sartorius muscles are plotted versus $[Na]_i$. Curves are labeled with the important experimental condition; the lower dotted line was obtained in the presence of both 5 m*M* Ba and 10^{-4} *M* ouabain. From R. A. Sjodin and O. Ortiz, *J. Gen. Physiol.* **66,** 269 (1975).

and K pumps and to approximately the same degree. To within the experimental errors involved, the results support a coupled Na-K pump with fairly rigid coupling of the Na and K fluxes under normal conditions. It is of interest to consider the coupling ratio, r, at high values of $[K]_o$, where the kinetics of the pumps show saturation and pumped fluxes are maximal. From Figures 6.2 and 6.3, V_{max} for the K pump is about 40 μmoles/g·hr or 26 pmoles/cm·^{2}sec. For a value of $[Na]_i$ = 40 mM, which is comparable to that holding in the measurements of the K pump rate, the maximal rate constant for the Na pump is 2.5 hr^{-1}, giving a maximal Na pump rate of 40 pmoles/cm·^{2}sec. The value calculated for r from maximal pumped flux data is thus 40/26 = 1.5, agreeing with the other methods used to estimate the coupling ratio. Provided that $[Na]_i$ is above a value of about 8 mM, the Na pumped flux varies linearly with increasing values of $[Na]_i$, the Na efflux showing a single rate constant under these conditions. As seen in Figure 6.4, pumped K influx also varies linearly with $[Na]_i$, so that for any particular value of $[K]_o$, the coupling ratio, r, remains constant as $[Na]_i$ is changed. To illustrate this fact, recall that the K flux data for Figure 6.4 was obtained with $[K]_o = 5mM$. With flux expressed in units of μmole/g·hr and $[Na]_i$ in units of μmole/g, the pumped K influx is given by $m_K = 0.55\ [Na]_i$ from the slope to the line marked 5 mM Ba. Obtaining the rate constant for the Na pump with $[K]_o = 5$ mM and $[Na]_o = 105$ mM (Figure 5.4), $m_{Na} = 0.80$ $[Na]_i$, which gives a constant coupling ratio equal to $0.80/0.55 = 1.46$.

THE DEPENDENCE OF THE SODIUM-POTASSIUM PUMP COUPLING RATIO ON P_K

As previously discussed, determinations of r depend upon assumptions regarding the value of P_K during pumping. Barium ions were useful in reducing P_K to isolate the pumped component of K influx. By comparing the Na pump rates measured normally with the K pump rates measured in the presence of Ba ions, estimates of r around 1.5 were obtained, which agree fairly well with estimates made in the absence of Ba ions, assuming that P_K does not change very much when the Na pump rate increases. In the absence of Ba ions, if P_K increases during pumping, the value of r increases and a greater fraction of the K influx is electrically coupled. Conversely, one might expect the value of r to decrease if P_K decreases. There is no experimental evidence bearing on the first point. The experiments in the presence of Ba ions, however, provide direct evidence for a decline in the value of r as P_K is reduced.[6] As P_K becomes reduced to low values, the Na pump could attempt to maintain the same pumping rate, adjusting K influx by the passive means of increased hyperpolarization. This kind of adjustment is slight when P_K is reduced by application of Ba ions, as additional hyperpolarization amounts at most to a few millivolts and the driving force for inward K movement remains about the same. Another possible adjustment is to decrease the pumping rate for Na or increase the pumping rate for K, or both, all of which would result in a decline in r. This is the kind of adjustment that actually occurs under these circumstances. The decline in Na pumping rate is small, about 5 percent, whereas the K pumping rate increases by about a factor of 1.5, the combined effect of which is to decrease the value of r to 1.1.

The Na:K pump in skeletal muscle fibers appears to have the capacity to alter its coupling ratio in response to changes in P_K. This fact alone makes it difficult to arrive at exact values for r under conditions in which P_K might change. Nevertheless, there are a variety of methods that give values around 1.5 for r. Combining kinetic data for the K pump measured in the presence of Ba ions with Na pump data measured normally in the absence of Ba gives estimates of 1.5 for r. If the normal K pump operates at 75 percent of the rate observed in Ba, however, estimates by the latter method are increased to about 2. It must be admitted, therefore, that uncertainties still exist about the estimation of r. Its value is the most unambiguous in the presence of Ba ions, where it is around 1.1, and it can be concluded with fair certainty that r can decrease to values close to unity and that the value of r decreases when P_K is lowered. The physiological advantage of such a mechanism would be that more K ions would be pumped inwardly if the possibility for K accumulation by electrogenic means were reduced. This could be part of a more general K-retention mechanism for muscle fibers.

The biophysical basis for a control mechanism of this sort remains obscure at present. As the membrane potential need not change much in this process, the pump sites themselves appear compelled to sense the number of open resting K flux channels. The implication is that pump sites and K channels must be close to one another and may be located in the same molecular machinery.

ACTIVATION OF THE SODIUM PUMP BY FOREIGN CATIONS AND THEIR INWARD TRANSPORT BY MUSCLE MEMBRANE

As is the case with other cells, cations other than K can activate the Na pump in muscle fibers. The foreign cations Rb, Cs, Li, and NH_4 have been reported to activate the Na pump in sartorius muscle fibers.[5,14,20-22] These results are not surprising in view of the fact that all of these cations have been reported to activate membrane ($Na^+ + K^+$)-ATPase, which undoubtedly forms the molecular basis for the Na:K pump.[23] The observed differences from activation by K ions are quantitative rather than qualitative and can be explained on the basis of differences in rates and cation affinities. From the standpoint of pump cation coupling, however, some interesting differences are apparent, due to the lower permeability for some of the cations in the given series.

A lower membrane permeability to the activating cation means that, for the same degree of pump stimulation, a greater fraction of the inward movement of the activating cation will be chemically coupled through the pump rather than electrically coupled through ionic channels. This is especially evident for Rb and Cs, where the lower values of P give rise to a higher membrane resistance during recovery. This results in greater hyperpolarization from the electrogenic effect of the pump. For Rb, the hyperpolarization is about 15 mV greater than for K[5], and for Cs, the hyperpolarization exceeds that for K by about 20 mV at the peak.[14]

In the case of pump activation by Rb ions, the observed changes in membrane potential cannot account for more than 10 percent of the Rb uptake. Thus about 90 percent of the Rb entry must be chemically coupled via the pump to Na extrusion. Even more inward Cs movement is chemically coupled, as P_{Cs} measured in strophanthidin-containing solutions appears to be no greater than P_{Na}.[14] The fact that Cs ions depolarize muscle fiber membrane to a greater extent than do Na ions is due to a reduction in P_K that occurs in the presence of Cs ions.[24] Within the limits of experimental error, all of the strophanthidin-sensitive Cs uptake into muscle cells is chemically coupled to Na extrusion. The behavior of the Na pump when activated by either Rb or Cs ions resembles the behavior observed when the pump is activated by K ions in the presence of Ba ions. Barium ions reduce P_K to approximately the value of P_{Rb}.[5,6] As the permeability of the activating cation becomes lower, the pump coupling ratio for Na and that activating cation appears to closely approach 1:1 in the case of skeletal muscle fibers from the frog.

ABNORMALITIES IN PUMP COUPLING DUE TO ALTERED METABOLISM

In addition to the normally occurring Na:K exchange, the Na pump can catalyze other modes of cation exchange. Activation by and transport of certain foreign cations has just been discussed. In addition, as discussed in Chapter 5, skeletal muscle fibers show an Na:Na exchange mode qualitatively similar to that shown by red blood cells.[25] As in red cells, a condition favoring the occurrence of this mode of pump coupling is normally the removal of external K ions.[26] Besides being inhibited by high external K levels, which convert Na:Na exchange to Na:K exchange, the Na:Na exchange mode of the pump has a strong relation to cellular metabolism. In red blood cells, there is evidence that the component of Na:Na exchange increases with decreasing levels of ATP. High levels of ADP also promote the Na:Na exchange mode of the Na pump in red blood cells.[27]

Evidence for similar mechanisms in excitable cells first appeared in studies on squid giant axons, where it was observed that cyanide renders the pump insensitive to external K but dependent on external Na.[28] A ouabain-sensitive Na influx in partially poisoned squid axons has also been observed.[29] It was subsequently found that the main metabolic alteration underlying enhanced Na:Na exchange in squid giant axons is an elevated intracellular ADP level and, hence, a reduced ATP/ADP ratio.[30,31] The Na:Na exchange catalyzed by the pump was unaffected by changes in internal P_i but was increased by elevating the level of internal ADP. The increase in the amount of Na:Na exchange was at the expense of a diminution of the Na:K exchange normally observed.

Further studies on skeletal muscle fibers involve the measurement of pumped Na and K fluxes in the presence of various metabolic inhibitors that alter metabolic parameters. Levels of Na pumping and modes of pump coupling show a strong dependence on metabolic conditions in the fibers.[32,33] One metabolic inhibitor that has been extensively studied in muscle fibers is 2, 4,-dinitrofluorobenzene (DNFB). Muscles treated with DNFB (0.38 m*M*) for 30 minutes show ATP levels depleted to about one-third of control values and ADP levels modestly augmented, resulting in a diminution of the ATP/ADP ratio to about one-fourth of control values.[33] An accompanying increase in Na efflux occurs in Na-enriched muscle fibers treated with DNFB. The increased Na efflux is inhibited by strophanthidin and occurs in the presence or absence of external K ions. The rise in Na efflux occasioned by DNFB thus reflects an increase in Na pump activity that does not require coupling to K influx. The increase in Na pumping rate is transient, reaching a maximum value in about 30 minutes and then declining. These effects seem to be definitely related to the altered ATP and ADP levels. The evidence for this is that a chemically unrelated inhibitor, iodoacetamide (IAM), produces similar effects on both ATP and ADP levels and also on Na pumping rate.[33] The transient nature of the response of the Na pump to metabolic inhibitors remains unexplained. The behavior is intimately related to the ATP and ADP levels, however, since the maximum Na efflux is attained much later with IAM than with DNFB and IAM requires a correspondingly longer period to lower the ATP/ADP ratio.

The fact that the Na pumping rate is enhanced by these metabolic inhibitors in the absence of external K ions strengthens the argument that the Na flux enhancement is not in the K-coupled component of the Na pump. This raises the question of the kinds of pump coupling that can occur in metabolically inhibited states. The available evidence suggests that with DNFB and IAM, the increased pump activity is an Na:Na exchange. When external Na is replaced with Li, the stimulation of the Na pump by DNFB is not seen.[33] The same workers found an increased Na influx into metabolically poisoned muscles.

The next question that arises is that of the fate of Na:Na coupled pump operation during metabolic inhibition. Given that an Na:Na exchange mediated by the pump is generated by DNFB treatment, is this at the expense of, or in addition to, the normal Na:K coupling of the pump? The magnitude of Na efflux enhancement by DNFB is greater in the absence of external K than in its presence. Three main possible reasons for this have been emphasized by Kennedy and DeWeer[33]: 1) the metabolic poisons are simply more effective in the absence of external K, 2) metabolic poisoning may produce a fixed amount of Na:Na exchange and simultaneously inhibit existing Na:K exchange, or 3) external K may inhibit Na:Na exchange so that more of it is seen in the absence than in the presence of K. As there is no direct way to further fractionate the components of Na efflux under these conditions, the evidence must come from influx measurements.

Measurements of strophanthidin-sensitive K influx showed that DNFB treatment had little, if any, effect on Na:K coupled pumping.[34] If Na:Na exchange is induced at the expense of Na:K exchange, DNFB should produce a reduction in the strophanthidin-sensitive K influx. The conclusion that Na:K coupling is not affected by DNFB is based, however, on the assumption that K inward pumping and K-activation of Na pumping reflect the same process, and that the magnitude of the K-activated Na efflux can be inferred from the magnitude of the strophanthidin-sensitive K influx. Evidence that this need not always be the case is discussed soon. In the case of DNFB the assumption is probably valid, as DNFB failed to affect either the ouabain-sensitive membrane potential or the membrane resistance. As the Na pump is strongly electrogenic in muscle fibers under these conditions, a change in Na:K coupling should be made manifest as a change in the ouabain-sensitive component of the membrane current.

Though DNFB treatment apparently does not affect Na:K exchange by the pump very much, external K does inhibit the Na:Na exchange induced by DNFB.[35] The DNFB-treated muscles behave normally in this regard because K_o also inhibits strophanthidin-sensitive Na influx in untreated muscle fibers. The following picture emerges on the basis of these results. In untreated muscles in K-free media, an electrically silent Na pumping occurs due to Na:Na exchange. Coupled Na:K pumping occurs both in addition to, and at the expense of Na:Na exchange as $[K]_o$ is increased. In DNFB-treated muscles, Na:Na exchange is induced by the ATP and ADP level changes occurring even in the presence of external K. The Na:Na exchange so induced is in addition to, and *not* at the expense of, the existing Na:K exchange level. As K_o inhibits Na:Na exchange, however, less exchange is induced by DNFB in the presence than in the absence of external K. Of the previously listed three possibilities, experiments indicate that the third possibility is the correct one.

The question arises as to the universality of these results: Do other metabolic inhibitors act in the same way? In single muscle fibers of the crab *Maia squinado*, application of the inhibitor atractylate causes a transient stimulation of the Na pump before the onset of pump inhibition.[36] Atractylate retards the movements of ATP and ADP across the mitochondrial membrane and could lead to the same kinds of changes in sarcoplasmic ATP and ADP levels observed with DNFB in frog muscle.

There is some evidence that IAM acts in a manner similar to that of DNFB.[33] This is not true for all inhibitors, however, as some notable differences are observed with azide. The addition of azide increases the Na efflux in frog muscle fibers that have been previously enriched with Na, but has no effect on the Na efflux in fresh muscle fibers with normal values of $[Na]_i$.[32] It is not known whether or not the lack of effect of azide on Na efflux from fresh muscles represents a difference from DNFB action, as the experiments with DNFB were evidently all performed on muscles with previously elevated Na contents.[33-35] Like DNFB, azide acts to produce a transient increase in the ouabain-sensitive Na efflux that is due to Na:Na exchange. It may be that an elevated value of $[Na]_i$ is a necessary catalyst for the induction of Na:Na exchange by metabolic inhibitors.

Also like DNFB, the azide-induced Na:Na exchange is in addition to, and not at the expense of the K-activated portion of Na pumping. Muscles undergo a normal net extrusion of Na in K-containing media in the presence of azide, and addition of K to K-free media brings about normal increases in ouabain-sensitive Na efflux as determined with ^{22}Na. The most striking difference between the results with azide and those with DNFB is that azide almost completely abolishes the ouabain-sensitive K influx in muscle fibers, even though the normal activating effect of K_o on the Na pump is preserved.[32] Thus in addition to inducing Na:Na exchange, azide uncouples the Na:K pump in muscle fibers so that a fairly normal K activation occurs in the absence of much inward pumping of K ions. Thus the activating effect of K ions on the Na pump and the inward active transport of K ions are processes separable by experimental operations.

The difference between the action of azide and DNFB on Na and K transport in muscle fibers is unknown as to mechanism. The inhibitor DNFB inhibits all ATP generation in frog skeletal muscle. It also inactivates creatine phosphokinase and interferes with oxidative phosphorylation and glycolysis.[33] Azide is an uncoupler of oxidative phosphorylation but also stimulates ATPase in muscle, the stimulation being followed by inhibition.[37] The biochemical basis for the action of azide on Na:K transport in muscle must await further experimentation. Unlike DNFB, azide depolarizes frog skeletal muscle fibers.[32,38] The membrane depolarization could be responsible for some of the differences observed between DNFB and azide action on Na-K transport in muscle. The depolarization observed with azide amounts to some 30 mV. Energetically, it is easier for the depolarized membrane to extrude Na ions and more difficult for the depolarized membrane to pump K ions inwardly. The combination of an energetics change and altered ATP and ADP levels may be responsible for the Na:K pump uncoupling observed with azide.

Inhibitors of metabolism continue to offer means to learn more about the biochemistry of Na and K transport in muscle fibers. An understanding of exactly how ATP and ADP levels act in the control of cation coupling of the Na pump in muscle appears somewhat remote at present, though it is clear that they must act in stabilizing conformations of the transport enzyme complex that are responsible for the different modes of Na and K transport.

Using a somewhat different approach, Erlij and Grinstein conclude that azide application leads to an actual increase in the number of Na-pump sites in the fiber membrane.[39] Ouabain was first applied to inhibit the Na pump. The fibers were then washed in ouabain-free Ringer's solution for 1 hour. No recovery of the pump was noted after this period. This finding favors the assumption that ouabain was still attached to existing pump sites. At this time, addition of azide caused a substantial increase in Na efflux that was subsequently blocked by readdition of ouabain. Control experiments indicated that azide did not simply act by reversing the ouabain inhibition. It was not shown whether or not the azide-induced increase in Na efflux was due to an Na:Na exchange, though this seems likely in view of the other work cited.

It thus seems possible that metabolically induced Na:Na exchange actually involves the creation of a new population of pump sites. The new pump sites do not occur as a consequence of the conversion of existing active pump sites (see previous discussion). They must arise, therefore, from conversion of existing precursor sites. A picture emerges of a muscle fiber membrane with a fairly labile distribution of Na-pumping sites in different conformations, including an inactive or precursor conformation. The various coupling modes of the pump can then be explained as differing proportions of the several conformations, and the induction of new pump sites can be explained as the conversion of a precursor conformation to one of the active conformations. If this picture proves to have general validity, the investigation of how the various conformations are brought under metabolic control clearly will form an exciting addition to existing knowledge of cation transport in muscle fibers and other cell types as well.

HORMONAL REGULATION OF SODIUM:POTASSIUM TRANSPORT

In addition to the ionic and metabolic regulation of the Na:K pump, three different hormones have been reported to have significant effects on Na and K transport in skeletal muscle fibers. These hormones are insulin, adrenaline, and thyroid hormone. The effects of these hormones and some possible mechanisms for action are briefly discussed.

In rat extensor digitorum longus muscle fibers, insulin produces membrane hyperpolarization and decreases both K efflux and influx.[40-42] The hyperpolarizing effect of insulin has been confirmed in frog muscle fibers as well.[43] This effect is blocked by ouabain, indicating some relation to the Na pump. Insulin also stimulates the Na:K pump in skeletal muscle fibers, as evidenced by an enhancement of ouabain-sensitive Na efflux and K influx determined with radioactive tracers.[44-47] It should be noted that the stimulatory effect of insulin on ^{42}K uptake was observed in rat soleus muscle,[46] as the effect differs from the observations of Zierler previously cited in the case of extensor digitorum longus muscles from the rat.[40-42]

There are conflicting reports on the action of insulin on membrane Na:K-activated ATPase. Rogus et al. found no significant effect of insulin on Na:K ATPase activity from rat skeletal muscle,[48] while other workers report increases in the activity of this enzyme from rat and frog muscle.[49,50] Further, insulin has been reported to augment the rate of ^{3}H-ouabain binding.[51] These observations have been used to argue that insulin acts specifically on the Na:K pump to increase its rate, possibly by enhancing the affinity of pump sites on the inner membrane surface for Na ions.[45,47]

Another possible action of insulin is to actually multiply the number of pump sites in the muscle membrane. To demonstrate such an action, use is made of muscles first inhibited by ouabain and subsequently washed free of extracellular ouabain.[52] The washed muscles are found to still have ouabain bound to pump sites, as shown by the lack of pump activity. Exposure to insulin brings about a marked increase in ouabain-sensitive Na efflux, presumably due to the induction of new pump sites that require new ouabain molecules for inhibition.[52] That insulin simply acts by unstabilizing the ouabain-enzyme complex is ruled out by experiments in which insulin is initially added to fibers prior to ouabain treatment. An increase in Na efflux is observed that is completely blocked by added ouabain. Such muscles are then washed free of extracellular ouabain for one hour, as before. A second addition of insulin now fails to increase the Na efflux by any amount. If insulin acts to destabilize the ouabain-enzyme complex, the second addition of insulin should result in a substantial increase in Na efflux.

Similar experiments indicate that most (80 percent) of the sites induced by insulin originate from the same pool of sites as those that give rise to azide-induced pump sites. These experiments support the idea that a common pool of precursor sites gives rise upon appropriate activation to active pump sites that can be regulated by either insulin or the ATP and ADP levels prevailing in the fibers. A common metabolic pathway for these effects is suggested. The idea that one of insulin's actions is to alter ATP and ADP affinities at enzymatic sites is very attractive.

As the azide-induced pump sites are sites that engage in Na:Na exchange that is not physiologically useful, a remaining area for investigation is the kind of cation coupling involved at the insulin-induced pump sites. These experiments must be performed before it can be decided how much physiological insulin action can be ascribed to the actual generation of new pump sites.

Whatever the questions remaining as to mechanism, there is general agreement that insulin hyperpolarizes skeletal muscle cells and leads to both additional Na extrusion and K accumulation. In rat extensor digitorum longus muscle fibers, the main action of insulin is apparently to lower cation permeability and specifically to lower the P_{Na}/P_K ratio.[53] The hyperpolarizing action of insulin in this instance can be ascribed more to permeability changes than to direct action on the Na pump. The net Na and K content changes then result from decreased cation leaks at more or less constant pumping rates. In the other fiber types discussed, the action of insulin seems to involve a direct stimulating effect on the Na:K pump, whether by way of increased turnover rates, increased cation affinities, or the induction of new pump sites. In these instances, increased K accumulation can occur by the K transport portion of the pump cycle, by the electrogenic effect of an increased Na pump rate on passive K influx, or by both mechanisms.

The insulin effects described are specific ones in the sense that they are separate from the action of insulin in glucose metabolism and are also separate from the transport effects of other hormones. The above insulin effects on Na:K transport are not modified by the omission of glucose or the addition of phlorizin.[46] Also, blocking of the Na:K pump by various means does not interfere with the effect of insulin on glucose transport.[54] In addition, there is evidence that the above described insulin effects on transport do not involve a reduction in the cytoplasmic concentration of cyclic AMP.[52,55]

Two other hormones have been shown to have effects on Na:K transport in skeletal muscle fibers, namely, adrenaline and thyroid hormone.

Adrenaline has a markedly stimulating effect on ouabain-sensitive Na efflux and K influx in isolated muscles, as does noradrenaline.[56,57] Catecholamines have a hyperpolarizing effect on skeletal muscle membrane that is early in its onset, and due to a direct stimulation of the electrogenic Na pump.[57,58] The effects of the catecholamines are mediated by β-2 type adrenoceptors and probably involve a stimulation of adenylate cyclase. As this process differs entirely from the action mechanism for insulin, insulin and adrenaline effects on transport would be expected to be additive as observed.[55] The stimulating effect of adrenaline on the Na:K pump is more pronounced than that due to insulin. Adrenaline accelerates ^{3}H-ouabain binding but apparently does not affect the total specific ouabain binding, indicating that the number of pump sites is not increased.[51] This is in contrast to the situation with insulin, where total specific ^{3}H-ouabain binding is increased.[52]

Thyroid hormone increases Na and K exchange in rat skeletal muscle due to increased membrane permeability.[59] The thyroid hormones have no immediate effect on Na:K transport in the case of isolated muscles studied *in vitro.* In the case of altered thyroid states, however, muscles show progressive increases in ouabain-sensitive ^{42}K influx in proportion to the degree of thyroid function.[60] Thyroid hormone increases the number of pump sites as determined by studies of ^{3}H-ouabain binding, which presumably explains the action on K influx.

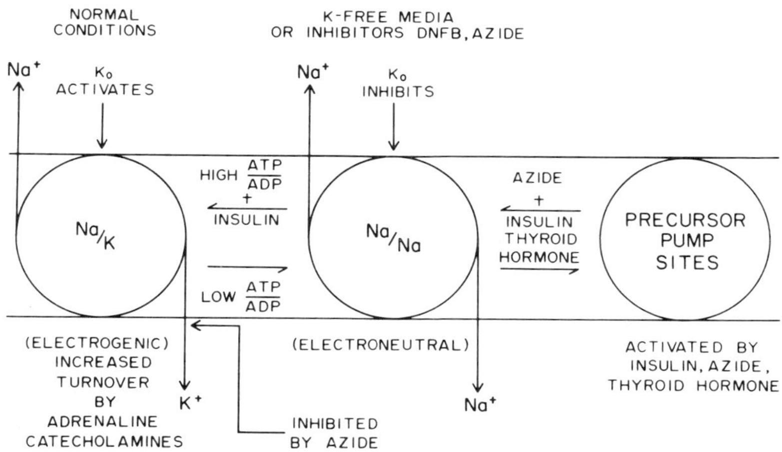

Figure 6.5. Summary of modes of coupling of the Na : K pump in skeletal muscle as influenced by metabolic factors and hormones.

Several aspects of the Na : K pump discussed in this chapter can be summarized as in Figure 6.5.

REFERENCES

1. R. P. Kernan, *Nature* (Lond.) **193,** 986 (1962).
2. A. S. Frumento, *Science* (N.Y.) **147,** 1442 (1965).
3. L. J. Mullins and M. Z. Awad, *J. Gen. Physiol.* **48,** 761 (1965).
4. S. R. Cross, R. D. Keynes, and R. Rybová, *J. Physiol.* **181,** 865 (1965).
5. R. H. Adrian and C. L. Slayman, *J. Physiol.* **184,** 970 (1966).
6. R. A. Sjodin and O. Ortiz, *J. Gen. Physiol.* **66,** 269 (1975).
7. E. J. Harris, *J. Gen Physiol.* **41,** 169 (1957).
8. R. D. Keynes, *J. Physiol.* **178,** 305 (1965).
9. R. A. Sjodin and L. Beaugé, *J. Gen Physiol.* **52,** 389 (1968).
10. D. Geduldig, *J. Physiol.* **194,** 521 (1968).
11. E. G. Henderson, *Life Sci.* **10,** 767 (1971).
12. R. A. Sjodin and L. A. Beaugé, *J. Gen Physiol.* **61,** 222 (1973).
13. R. A. Sjodin and A. Medici, *Nature* (Lond.) **255,** 632 (1975).

14. L. A. Beaugé and R. A. Sjodin, *J. Physiol.* **194,** 105 (1968).
15. R. D. Keynes and R. Rybová, *J. Physiol.* **168,** 58P (1963).
16. L. J. Mullins and K. Noda, *J. Gen. Physiol.* **47,** 117 (1963).
17. A. L. Hodgkin and P. Horowicz, *J. Physiol.* **148,** 127 (1959).
18. L. A. Beaugé, A. Medici, and R. A. Sjodin, *J. Physiol.* **228,** 1 (1973).
19. A. A. Lev, *Nature* (Lond.) **201,** 1132 (1964).
20. L. A. Beaugé and O. Ortiz, *J. Expl. Zool.* **174,** 309 (1970).
21. L. A. Beaugé, *J. Physiol.* **246,** 397 (1975).
22. L. A. Beaugé and R. A. Sjodin, *J. Gen. Physiol.* **52,** 408 (1968).
23. J. C. Skou, *Biochim. Biophys. Acta* **42,** 6 (1960).
24. R. A. Sjodin, *J. Gen. Physiol.* **42,** 983 (1959).
25. P. J. Garrahan and I. M. Glynn, *J. Physiol.* **192,** 189 (1967).
26. R. D. Keynes and R. A. Steinhardt, *J. Physiol.* **198,** 581 (1968).
27. I. M. Glynn and J. F. Hoffman, *J. Physiol.* **218,** 239 (1971).
28. P. C. Caldwell, A. L. Hodgkin, R. D. Keynes, and T. I. Shaw, *J. Physiol.* **152,** 591 (1960).
29. P. F. Baker, M. P. Blaustein, R. D. Keynes, J. Manil, T. I. Shaw, and R. A. Steinhardt, *J. Physiol.* **200,** 459 (1969).
30. P. DeWeer, *J. Gen. Physiol.* **56,** 583 (1970).
31. P. DeWeer, *Ann. N.Y. Acad. Sci.* **242,** 434 (1974).
32. L. A. Beaugé and R. A. Sjodin, *J. Physiol.* **263,** 383 (1976).
33. B. G. Kennedy and P. DeWeer, *J. Gen. Physiol.* **68,** 405 (1976).
34. B. G. Kennedy and P. DeWeer, *Nature* (Lond.) **268,** 165 (1977).
35. B. G. Kennedy and P. DeWeer, *Biophys. J.* **17,** 154a (1977).
36. E. E. Bittar, *J. Physiol.* **187,** 81 (1966).
37. E. Racker, *Mechanisms in Bioenergetics,* Academic Press, N.Y. (1965).
38. P. Horowicz and C. J. Gerber, *J. Gen. Physiol.* **48,** 515 (1965).
39. D. Erlij and S. Grinstein, *J. Physiol.* **259,** 33 (1976).
40. K. L. Zierler, *Am. J. Physiol.* **197,** 515 (1959).
41. K. L. Zierler, *Am. J. Physiol.* **197,** 524 (1959).
42. K. L. Zierler, *Am. J. Physiol.* **198,** 1066 (1960).
43. R. D. Moore and J. L. Rabovsky, *Am. J. Physiol.* **236,** 249 (1979).
44. R. Creese, *J. Physiol.* **197,** 255 (1968).

45. R. D. Moore, *J. Physiol.* **232,** 23 (1973).

46. T. Clausen and P. G. Kohn, *J. Physiol.* **265,** 19 (1977).

47. H. Kitasato, S. Sato, K. Murayama, and K. Nishio, *Jap. J. Physiol.* **30,** 115 (1980).

48. E. Rogus, T. Price, and K. L. Zierler, *J. Gen. Physiol.* **54,** 188 (1969).

49. B. P. Brodal, E. Jebens, V.Öy, and O. J. Iverson, *Nature* (Lond.) **249,** 41 (1974).

50. W. A. Gavryck, R. D. Moore, and R. C. Thompson, *J. Physiol.* **252,** 43 (1975).

51. T. Clausen and O. Hansen, *J. Physiol.* **270,** 415 (1977).

52. D. Erlij and S. Grinstein, *J. Physiol.* **259,** 13 (1976).

53. K. L. Zierler, Insulin, Ions and Membrane Potentials, in *Handbook of Physiology,* Section 7: *Endocrinology,* Williams and Wilkins, Baltimore (1972).

54. P. G. Kohn and T. Clausen, *Biochim. Biophys. Acta* **225,** 277 (1971).

55. J. A. Flatman and T. Clausen, *Nature* (Lond.) **281,** 580 (1979).

56. E. T. Hays, T. M. Dwyer, P. Horowicz, and J. G. Swift, *Am. J. Physiol.* **227,** 1340 (1974).

57. T. Clausen and J. A. Flatman, *J. Physiol.* **270,** 383 (1977).

58. N. Tashiro, *Br. J,. Pharmacol.* **48,** 121 (1973).

59. Y. Asano, *Experientia Suppl.* **32,** 199 (1978).

60. R. Biron, A. Burger, A. Chinet, T. Clausen, and R. Dubois-Ferrière, *J. Physiol.* **297,** 47 (1979).

7

Mechanisms of Anion Movements

Over 90 percent of all muscle chloride is extracellular, and this presents a number of difficulties in obtaining an accurate measurement of the remaining internal chloride ion concentration. Nevertheless, it is likely that chloride ions are passively distributed in skeletal muscle.

In recent measurements using chloride-sensitive microelectrodes,[1] E_m and E_{Cl} in frog sartorius muscle agreed within 1–3 mV over a wide range of membrane potential and intracellular Cl ion activity (A_{Cl_i}). When $[K]_o$ was 2.5 mM in a bicarbonate-free medium, A_{Cl_i} was 2.7 mM, agreeing with direct measurements of $[Cl]_i$, assuming an approximate value of 0.8 for the activity coefficient.

Nevertheless, some evidence does exist for a chloride pump. When P_{Cl} was reduced by placing the fibers in an acid solution, E_{Cl} moved to values more positive than E_m, which would indicate an intracellular Cl accumulation by a nonpassive mechanism. It was concluded that an active transport mechanism for Cl ions is present in skeletal muscle, but that it operates at a low rate usually obscured by large passive fluxes of Cl ions.

Despite the evidence for active Cl transport, the early suggestions that Cl ions are passively distributed in skeletal muscle[2,3] and that most of the Cl fluxes are also passive continue to be valid approximations.

ANION INTERACTIONS

Foreign anions slow the movements of Cl ions across the membrane of frog muscle fibers[3,4] in a manner similar to the action of foreign cations on the K fluxes in muscle.[5,6] It is noteworthy that the foreign anions do not affect cation exchanges across the muscle fiber membrane,[7] and this suggests two separate mechanisms. It is also noteworthy that the membrane resistance of muscle fibers is increased in the presence of foreign anions.[8]. This would be expected if foreign anions slowed the movement of Cl ions through Cl-specific channels, as Cl ions are the principal contributors to the muscle fiber membrane conductance. In order of increasing effect, the foreign anions Br^-, NO_3^-, I^-, ClO_4^-, and CN^- diminished Cl output when they were used to replace a portion of the extracellular Cl ions.[4] These anions have all been shown to be capable of penetrating into the fibers. The efflux of Cl ions is also impeded by the nonpenetrating anions sulphate, phosphate, and bicarbonate.

When muscle fibers are loaded with tracer Br ions, the rate of Br loss is also a function of the anion composition of the external medium. The rate of Br loss is slowed when external Cl is replaced with Br ions. These findings suggest that Br ions can use the Cl channel for transit and that entering anions interfere with anions exiting the channels, Br ions being more effective blockers than Cl ions. It could be postulated that the rate of exchange of the foreign anion with muscle Cl is limiting the rate of Cl output. For example, P_{Br} is less than P_{Cl}. As a result, Cl ion loss to a Br medium may be lessened because the Cl:Br exchange rate is lower than the Cl:Cl exchange rate. Two facts tend to rule out such an interpretation: first, Cl ion slowing occurs upon partial replacement of extracellular Cl for the foreign anion, and second, nonpermeating anions that do not exchange at all with internal Cl are no more effective than penetrating anions in blocking the Cl efflux.

There is some evidence, however, that part of the anion movement in muscle fibers represents an anion exchange and that part is due to a movement coupled to K ion movement.[4]. For example, the loss of Cl ions to solutions where Cl has been replaced by a nonpenetrating anion must represent an efflux that is limited to the combined net $Na^+ + K^+$ output. Actual measurements of the rates of K output to K-free NaCl or K-free $NaNO_3$ media showed that the rate of loss was about the same in each solution. Thus the rate of loss of the Cl that moves with K in KCl-loaded muscles appears not to be influenced by the foreign anions. In KCl-loaded muscles where Cl output in a NO_3 medium is compared with Cl output in a Cl medium, much more tracer Cl is lost in the Cl medium than in the NO_3 medium, indicating the presence of an extra Cl:Cl exchange in the Cl medium. There remain ambiguities in these interpretations until the possible role of the membrane potential in the various solutions has been deduced, as Cl movement is very sensitive to the electrochemical driving force for Cl ions. In the case of NO_3^- replacement, at least, the effects must be due to changes in P_{Cl} because the membrane potential in NO_3 Ringer's solution does not differ from that in Cl Ringer's solution.[3]

Spurway[9,10] has suggested that an externally directed membrane site plays a key role in the regulation of Cl ion permeability. Anion influx, too, is subject to interaction effects. However, only the external anion is important, not the internal anion. This is shown by preloading the muscle fiber with an anion of the retarding type, say Br ion. Further Cl entry is inhibited only if Br ions are also present on the outside. In a normal Cl medium, internal Br does not inhibit Cl entry. These facts suggest that it is the presence of anions at externally directed membrane sites that determines anion permeability. The characteristics of these sites form the subject of the next section.

ANIONIC MECHANISMS AS INFLUENCED BY pH

The sensitivity of the chloride conductance of frog muscle membrane to changes in external pH has been extensively studied by Hutter and Warner.[11-13] The pH effects on chloride conductance were evidenced by changes in the membrane potential and in the voltage deflections due to 600 msec constant current hyperpolarizing pulses. When normal Ringer's solution (pH 7.4) was made acid (pH 5.0), the resting potential increased from a value of -85 to -87 mV and the magnitude of the electrotonic potential became greater, indicating a drop in the membrane conductance. Increasing the external pH to 9.8 had opposite effects: E_m changed to about -81 mV, and the electrotonic potentials decreased to about half the size that they were in the acid solution. As $G_m \alpha 1/E_m^2$, the conductance changed four-fold when the pH went from 5.0 to 9.8.

These changes in E_m and G_m could be due to changes in P_K, P_{Cl}, or both. When external Cl was replaced by the relatively nonpermeating methylsulfate ion ($CH_3SO_4^-$), the effects of pH on the membrane conductance were no longer observed. When external pH was changed as before, little difference in the amplitude of the electrotonic potentials was observed. The membrane conductance is thus not sensitive to pH when the K ion is the only major current-carrying ion, and the previously described differences in the membrane conductance upon varying the pH must be due to changes in Cl ion conductance.

When pH was changed stepwise, it was possible to determine the quantitative relation between G_{Cl} and the pH. The relation obtained is shown in Figure 7.1, where the relative membrane conductance is plotted against pH. The contribution due to G_{Cl} at each pH is the difference between values in NaCl (open symbols) and values in $NaCH_3SO_4$ (filled symbols). The chloride conductance rises with increasing pH along a sigmoidal curve that is steepest at around pH 7.

As the pH changes could alter charged ionic groups in the membrane, and as Ca ions are known to influence charged membrane groups, experiments were performed to investigate the possible influence of Ca ions on the pH effects observed. It was found that wide variations in the Ca concentration had little influence on the magnitude of the conductance changes observed when the external pH was changed.

The possibility that the effects of pH on Cl conductance are mediated by changes in the states of ionized groups at the fiber membrane surface was also examined by Hutter and Warner.[11] A membrane in which fixed charges regulate ionic movements would be expected to attract ions of charge opposite to that of the fixed charge, and to exclude ions of the same sign as the fixed charge. The binding of protons by charged membrane groups would, on a simple basis, neutralize fixed negative charges and, hence, increase positivity or, alternatively, protonation of a neutral group would produce a fixed positive charge. Both events would, on a theoretical basis, be expected to increase anion permeability. Such behavior upon acidification has been observed in giant muscle fibers from the barnacle, where a sharp increase in Cl^- conductance is observed when the pH is lowered to values below pH 5.[14] Above pH 5, Cl^- conductance was insensitive to the pH. These effects could be explained by a simple fixed charge model in which protonation occurs below pH 5. The results with skeletal muscle fibers from the frog, however, do not conform to this simple model. Accordingly, a modified model was proposed by Hutter and Warner[11] to account for their results. The following reactions were suggested:

$$R^- + Ca^{2+} \rightleftharpoons RCa^+$$

$$H^+ + R^- \rightleftharpoons RH$$

The first reaction with Ca^{2+} is visualized as producing the fixed charge RCa^+ that is required for high Cl^- permeability. The second reaction involving protonation converts R^- to a form RH that cannot react to form the positive fixed charge required for high chloride permeability.

The major weakness in this model is that no competition between H^+ and Ca^{2+} was observed over a wide concentration range. If the affinity for Ca^{2+}

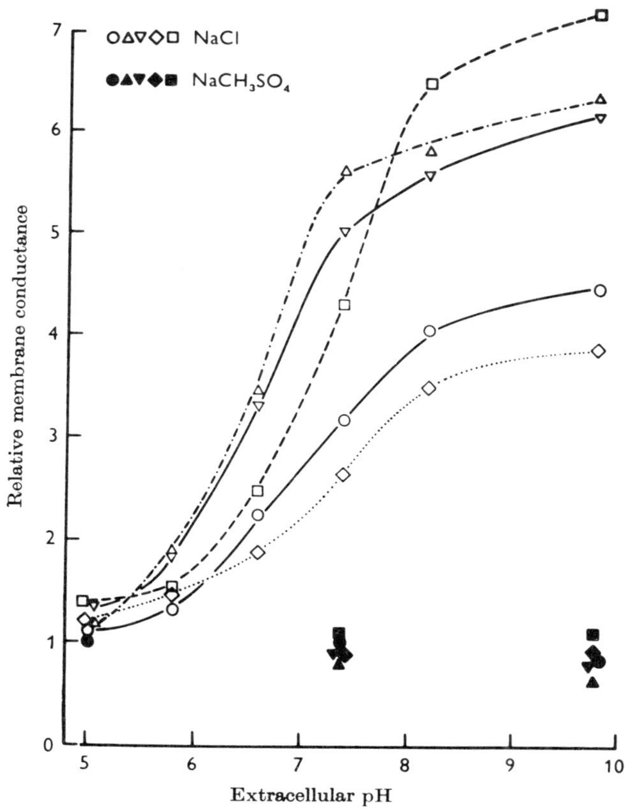

Figure 7.1. Dependence of membrane conductance on pH. Results from four sartorius fibers and one toe muscle fiber (□) of different frogs. Open symbols: muscles in chloride solution. Filled symbols: muscles in methylsulfate solution. Ordinate: relative membrane conductance in terms of conductance in methylsulfate solution at pH 5.0. From O. F. Hutter and A. E. Warner, *J. Physiol.* **189,** 403 (1967).

were very high, however, reducing $[Ca^{2+}]_o$ might be without effect in alkaline solutions. In acid solutions, on the other hand, lowering $[Ca^{2+}]_o$ might influence the density of charged membrane groups, but in a way that could not be detected by conductance measurements alone, as G_{Cl} is already low compared to G_K, and further reductions in G_{Cl} would be hard to detect from measurements of G_m alone. That this explanation does not suffice, however, is indicated by experiments with ^{36}Cl, in which it was shown that no change in Cl efflux occurs at any value of pH when Ca^{2+} is removed externally.[12] A further argument against the above modified fixed-charge model is that foreign divalent cations such as Cu^{2+} and Zn^{2+} also decrease chloride permeability.

These facts led Hutter and Warner to propose a more complicated model based upon a macromolecule, P, and an anion species, A, for which the anion affinity of P is modified by H^+. The anion affinity of certain proteins is known to increase upon protonation,[15] providing a physical basis for the model, which can be described as follows:

$$\mathrm{P + H \rightleftharpoons PH} \qquad \mathrm{PH + A \rightleftharpoons APH}$$

$$\mathrm{P + A \rightleftharpoons AP} \qquad \mathrm{AP + H \rightleftharpoons APH}$$

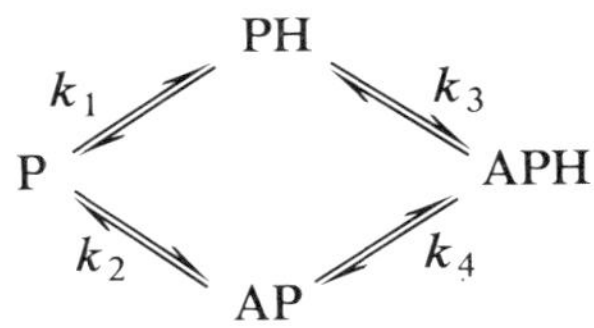

in which $k_3/k_2 = k_4/k_1 > 1$.

The following kinetic equations describe this system:

$$k_1\,[\mathrm{P}][\mathrm{H}] = [\mathrm{PH}] \tag{76}$$

$$k_2\,[\mathrm{P}][\mathrm{A}] = [\mathrm{AP}] \tag{77}$$

$$k_3\,[\mathrm{PH}][\mathrm{A}] = [\mathrm{APH}] \tag{78}$$

$$k_4\,[\mathrm{AP}][\mathrm{H}] = [\mathrm{APH}] \tag{79}$$

The fraction of sites occupied by anions is given by

$$\bar{\nu}_{\mathrm{A}} = \frac{k_2[\mathrm{A}] + k_1k_3[\mathrm{H}][\mathrm{A}]}{1 + k_1[\mathrm{H}] + k_2[\mathrm{A}] + k_1k_3[\mathrm{H}][\mathrm{A}]} \tag{80}$$

and

$$\bar{\nu}_{\mathrm{A}} = \frac{k'_{\mathrm{A}}[\mathrm{A}]}{1 + k'_{\mathrm{A}}[\mathrm{A}]} \tag{81}$$

where k'_A is a pH-dependent anion affinity coefficient given by

$$k'_A = \frac{k_2 + k_1 k_3[H]}{1 + k_1[H]} \tag{82}$$

The anion permeability would then depend upon two factors, the partition coefficient for anions $\bar{\nu}_A/[A]$ and the rate of dissociation of the anion-site complex. Accordingly, P_{Cl} would depend upon the product $(\bar{\nu}_A/[A])(1/k'_A)$ where A represents Cl. Thus

$$P_{Cl} = K(1 - \bar{\nu}_{Cl}) \tag{83}$$

with $\bar{\nu}_{Cl}$ given by equation (80) or (81), K being a proportionality constant. Using equation (82) for the pH dependence, a plot of P_{Cl} versus pH so calculated agrees well with the plot of the actual data presented in Figure 7.1.

The model just described appears to be adequate for results on skeletal muscle as they now exist. One might inquire as to the generality of the model and also ask how results on skeletal muscle fit in with the general picture on anion permeability that is emerging from studies of other cell types. It has already been mentioned that findings relative to the chloride conductance of giant muscle fibers from the barnacle conform to the simple amphoteric site model.[14] The chloride ion conductance of crayfish muscle fibers becomes reduced at high pH,[16,17] also agreeing with the simple model rather than with the more complicated model necessitated by the results on skeletal muscle fibers. Red blood cells, on the other hand, show the same type of dependence of chloride permeability on pH that is shown by skeletal muscle cells from pH 5 to 9.[18] Above pH 9, however, P_{Cl} falls steeply in red blood cells. Also, the steep part of the decline in P_{Cl} in acid solutions occurs at lower pH values in red cells than it does in skeletal muscle cells. Another difference in red cells is that external protons do not inhibit Cl ion transport unless sulphate ions are present. Apparently SO_4^{2-} and Cl^- compete for an external transport site and H^+ and SO_4^{2-} are cotransported to the inside, where protons exert their action on the membrane sites. Thus no single anion transport scheme is able to account for the great diversity of results obtained with different cell types.

In skeletal muscle cells, Cl ion transport makes a strong contribution to the membrane conductance, whereas in red cells, the large anion exchange fluxes do not contribute to membrane conductance.[19,20] Perhaps basically different mechanisms should be looked for in different cell types. It has recently been pointed out that a wide variety of inhibitor action on anion transport in red cells can be accounted for by isolation of a specific transport protein (Band 3 protein), whose amino groups are active in anion transport.[21] Incorporation of purified Band 3 protein into liposomes confers on this system anion permeability that has properties in common with anion permeability in red blood cells. This protein may have a different conformation in skeletal muscle cells, namely, a conformation giving rise to chloride conductance channels. In any event, anion transport in cells is clearly a complicated process with many more features to be worked out.

EXTENSION OF ANION TRANSPORT MODEL IN SKELETAL MUSCLE

Any successful model describing anion transport in skeletal muscle cells must be able to account for the monovalent anion interactions previously described, the ability of certain divalent foreign cations to inhibit Cl ion transport, and the inability of certain other divalent cations to influence Cl ion fluxes. The problem has been attacked both experimentally and theoretically by Hutter and Warner.[13] The effects of foreign cations were tested by first showing that upon changing the pH from 5.0 to 7.4, the preparation gave a large increase in Cl conductance as determined by ^{36}Cl efflux. The foreign cations were then added externally at pH 5.0 and they were allowed to equilibrate with the membrane sites. The pH was then changed to 7.4 and ^{36}Cl efflux was again determined. The results of experiments with Cu^{2+} and Zn^{2+} are illustrated in Figure 7.2. Both foreign cations clearly caused large reductions in ^{36}Cl efflux. Further experimentation at different concentrations of the inhibiting divalent cation indicated that Cu^{2+}, Zn^{2+}, and UO_2^{2+} all inhibited Cl efflux with approximately the same affinity as indicated by the data illustrated in Figure 7.3. The steep part of the curve describing inhibition occurs at a concentration of about 10^{-4} M, indicating an affinity for these cations three orders of magnitude less than that for hydrogen ions. By contrast, the cations Ni^{2+}, Co^{2+}, Pb^{2+}, Ca^{2+}, Ce^{3+}, and La^{3+} had no effect on Cl fluxes.

Monovalent anion interactions were further studied by measuring their influence on Cl flux at different pH values and also by measuring the influence of pH on the fluxes of the foreign monovalent anions. The external presence of I ions strongly suppressed the Cl efflux in alkaline solutions. Iodide ions inhibited Cl efflux to a much lesser degree in acid solutions. A corollary of

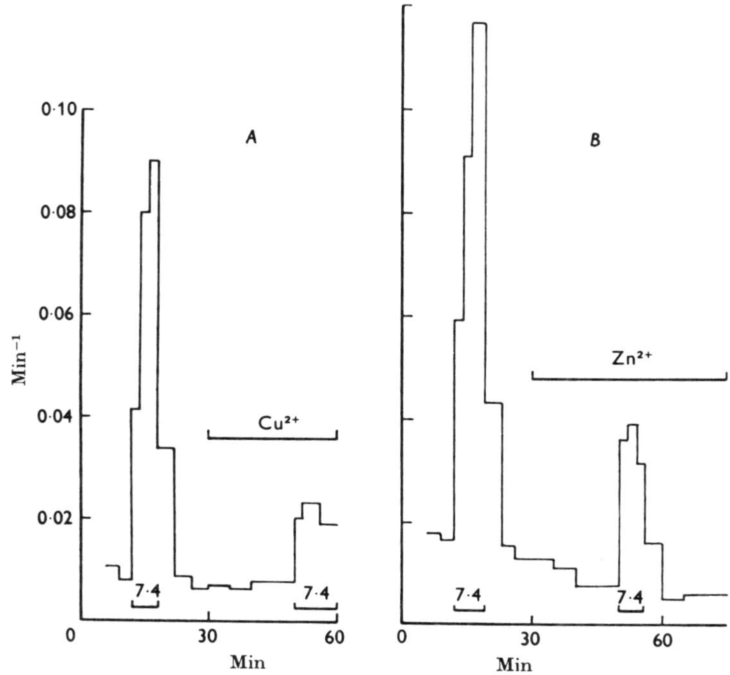

Figure 7.2. Influence of Cu^{2+} and Zn^{2+} on the chloride permeability of frog toe muscles. Ordinate: rate coefficient for efflux of ^{36}Cl. Abscissae: time from beginning of efflux into inactive wash solution at pH 5.0. pH transiently raised to 7.4 before and after addition of 5×10^{-4} *M* $CuCl_2$ (*A*) and 5×10^{-4} *M* $ZnCl_2$ (*B*). From O. F. Hutter and A. E. Warner, *J. Physiol.* **189,** 445 (1967).

these findings is that, in the presence of iodide ions, alkaline pH stimulates Cl flux to a much lesser extent. The effect of pH variation on the efflux of ^{82}Br and ^{131}I was also studied. When Br efflux was followed into a Br Ringer's solution, elevation of pH increased Br efflux but to a lesser degree than that observed for Cl efflux into normal Ringer's solution. The effect of pH on I ion efflux into I Ringer's solution was small. The absolute values of P_I and P_{Br} were lower than P_{Cl} in alkaline solutions, but in acid solutions the membrane discrimination between the anions was much less. In addition, the effect of external Cl on I ion efflux was determined. Changing from an I Ringer's solution to a Cl Ringer's solution increased the I ion efflux at both pH 5.0 and 9.8. However, in Cl Ringer's solution or in Ringer's solution with

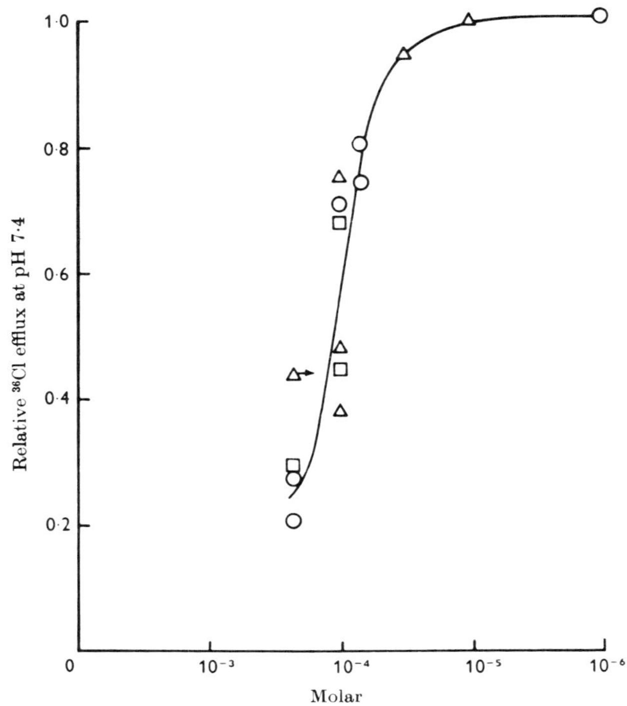

Figure 7.3. Decrease in ^{36}Cl efflux at pH 7.4 produced by various concentrations of $CuCl_2$ (O), $ZnCl_2$ (□), and $UO_2(CH_3COO)_2$ (Δ). From O. F. Hutter and A. E. Warner, *J. Physiol.* **189,** 445 (1967).

only 5 m*M* Cl substituted by I ions, the efflux of I was decreased slightly in alkaline solutions, an effect opposite in direction to the usual pH effects.

Many of these findings can be rationalized by an extension that Hutter and Warner made of their initial model. The extension is diagrammed below:

$$\begin{array}{ccccc} \text{PM} & \overset{k_5}{\rightleftharpoons} & \text{P} & \overset{k_1}{\rightleftharpoons} & \text{PH} \\ k_7 \updownarrow & & \updownarrow k_2 & & \updownarrow k_3 \\ \text{APM} & \underset{k_6}{\rightleftharpoons} & \text{AP} & \underset{k_4}{\rightleftharpoons} & \text{APH} \end{array}$$

The required additional reactions are

$$\text{P} + \text{M} \rightleftharpoons \text{PM} \quad \text{and} \quad \text{A} + \text{PM} \rightleftharpoons \text{APM}$$

where M refers to a cation and one bound cation promotes the binding of only one anion. The additional equations are

$$[\mathrm{PM}] = k_5[\mathrm{P}][\mathrm{M}] \tag{84}$$

and

$$[\mathrm{APM}] = k_7[\mathrm{A}][\mathrm{PM}] = k_5k_7[\mathrm{A}][\mathrm{P}][\mathrm{M}] \tag{85}$$

The permeability for the anion, as before, is assumed to be proportional to the product of the partition coefficient and the reciprocal of the anion affinity coefficient, which, mathematically, is equivalent to proportionality to the fraction of sites *not* occupied by anions:

$$P_\mathrm{A} = K(1 - \bar{\nu}_\mathrm{A}) = K\left[\frac{1 + k_1[\mathrm{H}] + k_5[\mathrm{M}]}{1 + k_2[\mathrm{A}] + k_1[\mathrm{H}] + k_5[\mathrm{M}] + k_1k_3[\mathrm{H}][\mathrm{A}] + k_5k_7[\mathrm{M}][\mathrm{A}]}\right] \tag{86}$$

where $k_3 > k_2$ and $k_7 > k_2$.

In understanding the Hutter-Warner model, it is important to realize that the permeability to an anion is determined by two factors, the number of sites occupied by anions per unit of applied concentration, that is, the partition coefficient, and the ease with which the anion complex dissociates to give free mobile anions. Cations exert an inhibitory effect, according to this model, by tightening the binding of the anion so that it dissociates from the sites at a slower rate. Thus the inhibitory divalent cations act essentially as if more hydrogen ions were bound. The failure of some divalent cations to influence chloride flux is easily explained by assuming very low affinities of the P and AP moieties for M. The slow Br movement in alkaline solution and the lesser pH sensitivity of Br permeability could be accounted for by assigning a higher value to the affinity of the unprotonated molecule for bromide than for chloride without a proportional increase in the affinity of the protonated molecule. To describe the iodide permeability, the affinities of the two forms have to be almost equal and higher than the chloride affinity of the protonated form.

To account for competitive anion effects such as the slowing of chloride movement by Br or I ions, an appropriate kinetic model is used to describe the anion competition in which the proportion of sites occupied by chloride ions at different pH in the presence of a foreign anion like iodide would be given by

$$\bar{v}_{\mathrm{Cl}} = \frac{k_2[\mathrm{Cl}] + k_1 k_3[\mathrm{H}][\mathrm{Cl}]}{1 + k_1[\mathrm{H}] + k_2[\mathrm{Cl}] + k_8[\mathrm{I}] + k_1 k_3[\mathrm{H}][\mathrm{Cl}] + k_1 k_9[\mathrm{H}][\mathrm{I}]} \qquad (87)$$

where k_8 and k_9 are the affinities of the unprotonated and protonated forms for iodide. The different degrees of slowing of the Cl efflux by iodide ions at different pH values can be interpreted in this way, assuming that only H and not I influences Cl binding.

The nearly neutral pK of the pH versus Cl^- conductance relation and the sensitivity to Cu^{2+} and Zn^{2+} and lack of sensitivity to Ca^{2+} have all been taken as signs pointing to a charged imidazole grouping as the chemical factor in the membrane that controls the chloride permeability in skeletal muscle fibers.

ACTIVE CHLORIDE TRANSPORT

Chloride ion activity data suggest that if a chloride pump is present in skeletal muscle fibers, it must be a fairly weak one. A weak chloride pump could best influence the Cl ion distribution under conditions in which the Cl permeability is low. For frog muscle fibers, acid solutions should favor the manifestation of a Cl ion pump, should one be present. Accumulation of fiber chloride above the value $[\mathrm{Cl}]_o \exp(EF/RT)$ does apparently occur in acid solutions.[11] Evidence for active Cl ion accumulation in excitable cells was first obtained in squid giant axons by Keynes.[22] In cardiac muscle where P_{Cl} is normally low, there is evidence for Cl ion accumulation under normal conditions.[23-25] Further evidence for a possible Cl pump in skeletal muscle fibers has recently been obtained.[26] However, a decline in ^{36}Cl efflux is associated with the observed Cl ion accumulation. More work is required to distinguish possible effects due to single-file diffusion and coupled ion transport from an actual Cl ion pump.

REFERENCES

1. T. B. Bolton and R. D. Vaughan-Jones, *J. Physiol.* **270,** 801 (1977).
2. P. J. Boyle and E. J. Conway, *J. Physiol.* **100,** 1 (1941).
3. R. H. Adrian, *J. Physiol.* **156,** 623 (1961).
4. E. J. Harris, *J. Physiol.* **141,** 351 (1958).

5. R. A. Sjodin, *J. Gen. Physiol.* **42,** 983 (1959).
6. R. A. Sjodin, *J. Gen. Physiol.* **44,** 929 (1961).
7. C. Edwards, E. J. Harris and K. Nishie, *J. Physiol.* **135,** 560 (1957).
8. S. M. Padsha, *J. Physiol.* **137,** 26P (1957).
9. N. C. Spurway, *J. Physiol.* **178,** 51P (1965).
10. N. C. Spurway, *J. Physiol.* **181,** 51P (1965).
11. O. F. Hutter and A. E. Warner, *J. Physiol.* **189,** 403 (1967).
12. O. F. Hutter and A. E. Warner, *J. Physiol.* **189,** 427 (1967).
13. O. F. Hutter and A. E. Warner, *J. Physiol.* **189,** 445 (1967).
14. S. Hagiwara, R. Gruener, H. Hayashi, H. Sakata, and A. D. Grinnell, *J. Gen. Physiol.* **52,** 773 (1968).
15. C. W. Carr, *Archs. Biochem. Biophys.* **46,** 417 (1953).
16. J. P. Ruben, L. Girardier, and H. Grundfest, *Biol. Bull. Mar. Biol. Lab., Woods Hole* **123,** 509 (1962).
17. W. DeMello and O. F. Hutter, *J. Physiol.* **183,** 11P (1966).
18. R. B. Gunn and O. Frohlich, *Ann. N.Y. Acad. Sci.* **341,** 384 (1980).
19. E. J. Harris and B. C. Pressman, *Nature* (Lond.) **216,** 918 (1967).
20. A. Scarpa, A. Cecchetto, and G. F. Azzone, *Biochim. Biophys. Acta* **219,** 179 (1970).
21. H. Passow, H. Fasold, E. M. Gärtner, B. Legrum, W. Ruffing, and L. Zaki, *Ann. N.Y. Acad. Sci.* **341,** 361 (1980).
22. R. D. Keynes, *J. Physiol.* **169,** 690 (1963).
23. E. Page, *J. Gen. Physiol.* **46,** 201 (1962).
24. E. E. Carmeliet and M. Janse, *Arch. Int. Physiol. Biochem.* **73,** 174 (1965).
25. O. F. Hutter and D. Noble, *J. Physiol.* **157,** 335 (1961).
26. D. D. Macchia and E. Page, *Fed. Proc.* **39,** 295 (1980).

8

Calcium Ion Movements

Calcium ions trigger the contraction of myofibrils. Hence, their movement in muscle fibers is of fundamental importance.* Unlike cardiac and smooth muscle fibers, the skeletal muscle fiber has sufficient internal calcium stores to make it relatively independent of the external calcium concentration. Thus skeletal muscle fibers in a Ca-free medium retain the ability to contract for long periods. Cardiac muscle, on the other hand, quickly loses the ability to contract upon being placed in a Ca-free medium.

If the external membranes of a skeletal muscle fiber were completely impermeable to Ca ions, a fiber would theoretically be able to contract indefinitely in a Ca-free environment. The Ca ions would be released from the sarcoplasmic reticulum (SR) to elicit contractions and then would simply be resequestered by the SR with no Ca loss to the exterior. However, as the skeletal muscle membrane appears to be about as permeable to Ca ions as to Na ions,[1] there will always exist an entrance and exit of Ca ions across the fiber membrane in accordance with the Ca gradients. For the Ca ion concentrations at rest, about 10^{-7} M inside and 10^{-3} M outside, there will be a strong tendency for Ca ions to enter the fibers, given an internal negativity of

*Calcium ion transport in the sarcoplasmic reticulum is such an important and highly investigated area that it is treated in detail in another volume in this series.

some 90 mV. This constant inward stream of Ca ions must, in some way, be extruded from the fibers to avoid a continual buildup of Ca within. Also, since contraction failure eventually does occur in a Ca-free medium, the fibers must rely to some extent on the entrance of Ca ions through Ca channels during the muscle action potential to maintain the ultimate balance of Ca ions that must exist on a longer time scale. The implication is that some Ca ions must be lost to the exterior during the cyclic events of Ca release and sequestration, which occur during contraction and relaxation, or, at least, that Ca ions in some special region must become depleted during the cycle, eventually becoming replenished by way of Ca entrance through Ca channels.

As there is not yet much agreement on models describing Ca movements in muscle, it may be helpful to consider the hypothetical sources of Ca ion movements in skeletal muscle fibers (Figure 8.1). Calcium movements across the surface membrane (SL = sarcolemma) are inferred from data on Ca ion permeability,[1] from observations on ^{45}Ca influx into muscle during excitation,[2] and from optical measurements of Ca levels in muscle fibers during excitation.[3] The storage of Ca in the terminal cisternae (TC) of the SR in resting muscle, and the subsequent passage of Ca ions into the myofibrillar space (MFS) during contraction have been demonstrated by radioautography.[4] The other processes are hypothetical and experimentally difficult to verify. The accumulation of Ca by SR fragments is well known. The mechanism for return of Ca to the TC is especially hypothetical. The exit step at the surface membrane most certainly involves a Ca pump and an Na:Ca transport process. It will be helpful to keep Figure 8.1 in mind during the ensuing discussion of Ca ion shifts.

CALCIUM ION FLUXES IN WHOLE MUSCLE

Measurement of Ca ion fluxes in whole muscle is complicated by the presence of several compartments to which Ca ions have access. For this reason, accurate Ca influx measurements using ^{45}Ca ions are practically impossible in whole sartorius muscle from the frog. In single giant muscle fibers from the barnacle, however, it has been possible to insert a glass scintillator probe intracellularly and make direct measurements of Ca ion influx.[2] The ^{45}Ca uptake curve determined by this technique is presented in Figure 8.2. With all corrections applied, and with about 5 m*M* EGTA inside the fibers, Ca influx at rest amounts to about 1.5 pmoles/cm^2 · sec in this preparation.

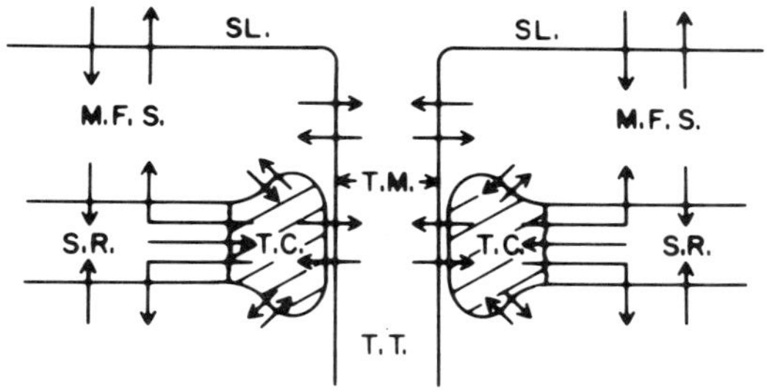

Figure 8.1. Diagram of some hypothetical sources of Ca ion movement in skeletal muscle fibers. SL., sarcolemma; M.F.S., myofilament space (sarcoplasm); T.M., tubular membrane; T.T., transverse tubule; T.C., terminal cisternae; S.R., sarcoplasmic reticulum.

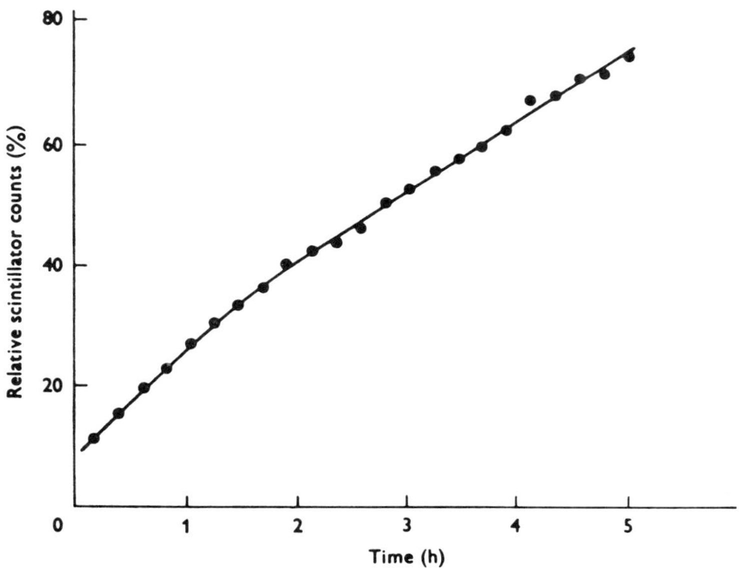

Figure 8.2. Relative Ca uptake versus time for a single muscle fiber from the barnacle, *Balanus nubilus.* Fiber diameter, 1.2 mm; temperature, 20°C. From C. C. Ashley, P. C. Caldwell, A. K. Campbell, T. J. Lea and D. G. Moisescu, Calcium Movements in Muscle, in *Symposium XXX of the Society for Experimental Biology,* C. Duncan, Ed., Cambridge University Press, Cambridge (1976), p. 397.

In frog sartorius muscle, the total net content of Ca has been followed as a function of $[Ca]_o$.[5] The results are illustrated in Figure 8.3. Considerable Ca was found in the connective tissue; therefore, measurements of the total whole muscle Ca content must be corrected for the connective tissue Ca in order to obtain total muscle fiber Ca. The results show that a net Ca gain occurs at values of $[Ca]_o$ above about 4 mM and that the normal resting fiber Ca content is about 2 μmoles/g wet weight.

In whole muscle, it is easier to study ^{45}Ca efflux than ^{45}Ca influx because the washout of fast compartmental fraction components can readily be seen and separated from the slow, presumably intracellular, compartment. A curve illustrating the washout of ^{45}Ca from frog sartorius muscle previously loaded with ^{45}Ca for a four hour period is shown in Figure 8.4. The rate constant for the slow component of ^{45}Ca washout is about 0.05 hr^{-1}. Shanes and Bianchi[6] used ^{45}Ca washout kinetics to separate the total Ca content (1.81 μmoles/g) into three fast and two slow components as follows:

Fast Compartment	Ca Content, μmole/g	Slow Compartment	Ca Content, μmole/g
Aqueous	0.16	Intracellular exchangeable	1.00
Connective tissue	0.16	Intracellular nonexchangeable	0.36
Fiber surface	0.13		

Applying the rate constant for slow ^{45}Ca loss to the total Ca in the exchangeable intracellular pool gives values of Ca flux that are about 50–100 times lower than values of the Na ion fluxes. Taking into account that the value of $[Ca]_o$ (2 mM) is much lower than the value of $[Na]_o$ (120 mM) and that total muscle Ca is in a steady state at this value of $[Ca]_o$, it can be calculated that the values of P_{Ca} and P_{Na} are roughly the same in frog muscle.

It has also been possible to measure Ca ion fluxes in single-twitch fibers from the frog's semitendinosus muscle.[7] The resting influx when $[Ca]_o = 1.0$ mM was 0.26 pmole/cm$^2\cdot$sec and ^{45}Ca efflux followed a single-exponential time course, indicating Ca loss from a single intracellular compartment. The first point after a few minutes of efflux was often above the best straight line that could be drawn through the other points on a semilogarithmic plot. This indicates the presence of a second compartment for Ca ions in each individual muscle fiber. Approximately one-half the muscle fiber Ca was ex-

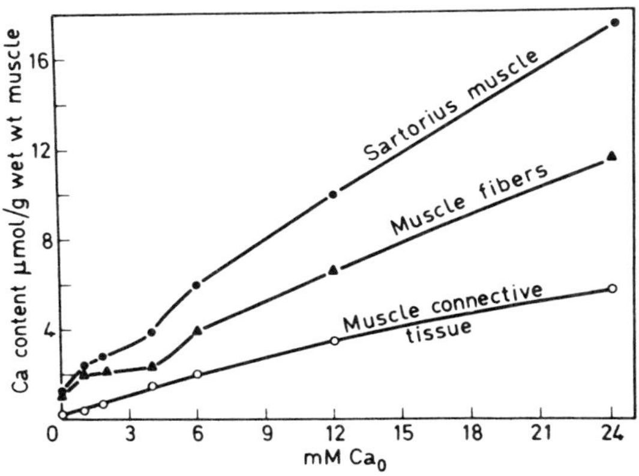

Figure 8.3. Calcium content of sartorius muscle, muscle fibers, and connective tissue as a function of external calcium concentration. From C. P. Bianchi, *Cell Calcium,* Appleton-Century-Crofts, New York (1968). [The data on which these curves are based were taken from D. L. Gilbert and W. O. Fenn, *J. Gen. Physiol.* **40,** 393 (1957).]

changeable. These single-fiber results essentially corroborate the results on whole muscle. The fact that ^{45}Ca washout shows a more pronounced multicompartmental nature in the case of whole muscle can be explained by extracellular diffusion and the presence of connective tissue.

CALCIUM FLUXES DURING MEMBRANE DEPOLARIZATION

As there exists only limited knowledge about the state of Ca ions in the various cellular compartments, it is difficult to extract more useful data from measurements of resting Ca ion movements in intact muscle fibers. Additional information can be obtained by studying the changes in Ca ion flux that occur when muscle fibers are depolarized by K or are electrically stimulated.

That Ca ions are the activators of the contractile machinery is now common knowledge in all disciplines studying muscle contraction. The reactions of Ca with the contractile elements and the relations between $[Ca^{2+}]_i$ and tension are beyond the scope of this monograph, which deals with transport only. Also, as has been already indicated, reactions of Ca with the SR are so important that they are dealt with in a separate monograph in this series.

When skeletal muscle fibers are depolarized with elevated external concentrations of K ions, they undergo long contractions, termed *contractures.*[8] When $[K]_o$ is maintained at an elevated level, the fibers relax after several

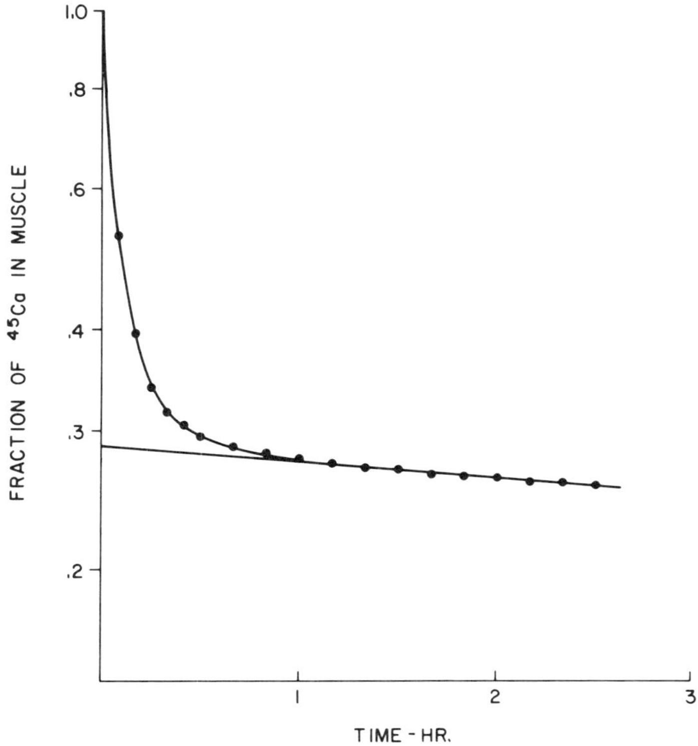

Figure 8.4. Washout of ^{45}Ca from frog sartorius muscle into Ringer's solution is plotted semilogarithmically against time. The muscle was previously loaded with ^{45}Ca for 4 hours. From R. A. Sjodin and S. Gold unpublished results.

seconds. Their ability to undergo subsequent contractures requires a period of repolarization prior to the initiating depolarization. As Ca ions are the activators of contraction, it was natural to look for an increased Ca ion entry during a K contracture. When Ca influx is measured during a K contracture, a transient increase in influx occurs that can be described by an S-shaped curve similar to that obtained when tension is measured.[1] When ^{45}Ca and K were added simultaneously, no increase in ^{45}Ca influx was observed. To obtain an increase in ^{45}Ca influx, the labeled Ca had to be added prior to addition of K. This was interpreted as evidence that ^{45}Ca ions had to first exchange with membrane Ca ions before penetrating into the fibers. When the chloride in Ringer's solution was replaced by nitrate ions, tension during the contracture was augmented, and increased ^{45}Ca entry occurred. Though a relationship between the entering Ca and the development of tension is certainly suggested, it is definitely not one of quantitative proportion. For example, if $[Ca^{2+}]_o$ is reduced to 0.1 m*M* in these studies, Ca entry is reduced by 90 percent while tension falls by only 30 percent. If Ca ions are removed from the external medium after a contracture, the ability to regain the capacity for a subsequent contracture upon repolarization is lost.[9] If Ca ions are not entirely removed from the external medium but are reduced in concentration, the rate of recovery during a period of repolarization is markedly slowed.[10]

Though studies of K contractures strongly suggest that contraction is initiated by entering Ca ions, difficulties are encountered when the same reasoning is applied to the increased Ca entry that occurs during an action potential. To be sure, an increased influx of Ca is associated with each impulse during electrical stimulation. In the case of frog skeletal muscle, the extra influx of Ca amounts to 0.20 pmole/cm^2 per impulse.[1] When external Cl is replaced by foreign anions that potentiate contraction, the Ca entry per impulse increases to 0.32 pmole/cm^2 (NO_3) and 1.35 pmoles/cm^2 (SCN). As in the case of contractures, the data suggest that Ca influx has a role in initiating contraction. However, Ca ion diffusion within the fiber is too slow to account for the rapid activation of the contractile elements that is observed. In addition, the amount of Ca entering per impulse gives too low an increase in $[Ca^{2+}]_i$ to activate all of the contractile elements within the fiber.

The results obtained with single frog muscle fibers are in essential agreement with those on whole muscle[7] except that no evidence for a membrane surface binding of Ca prior to entry was found. The Ca ion entry occurring when isotope was applied to the fiber at the onset of stimulation was compared with that measured when the fiber was allowed to equilibrate with isotope for some time prior to the commencement of stimulation. The two Ca ion entry rates per impulse were the same. In single fibers, therefore, there is no need to postulate that entering Ca ions must first equilibrate with a surface pool of membrane-bound Ca ions. If such surface binding occurs, it must be so rapid as to be indistinguishable from equilibration of an unstirred aqueous layer immediately surrounding the fiber. The longer delays seen in whole-muscle experiments must be due to longer diffusion times in the interstitial spaces surrounding the fibers.

Further evidence has been obtained that the Ca ions entering the fiber by crossing the surface membrane are not those that activate the contractile elements during a twitch. In EGTA buffered solutions in which $[Ca^{2+}]_o$ is lower than the internal free Ca ion concentration, normal twitches are still obtained.[11] Calcium deprivation in the external medium does cause K contractures to become smaller and shorter, suggesting at least some role for external Ca in the activation of contraction.[12] Further experiments, however, demonstrate conclusively that under conditions in which Ca influx is negligible in magnitude, depolarization still induces contraction.[13] In these experiments, the EGTA has undoubtedly entered into the T-system so that it is not likely that a Ca flux across the transverse tubuluar membrane is the signal for the SR to release Ca^{2+}. In frog skeletal muscle fibers at least, the role of Ca ions entering across the membrane during depolarization remains obscure. An increased Ca efflux during stimulation has also been reported in frog skeletal muscle,[14] but its significance remains unknown at present.

Similar studies have been performed on single striated giant muscle fibers from the barnacle, *Balanus nubilus.* As with resting fluxes, Ca fluxes during stimulation have been measured using ^{45}Ca ions and an internal scintillation probe.[2] As occurs in vertebrate skeletal muscle fibers, depolarization of the barnacle muscle fiber membrane triggers increased Ca uptake. During the application of constant current pulses of 200 msec duration at a rate of 4 pulses/sec, there was an increased Ca uptake of 2–3 pmoles/cm^2 per pulse when the average depolarization was 75 mV and $[Ca^{2+}]_o = 11.8$ m*M*. When the average depolarization was less (33 mV), the Ca uptake was less (0.73 pmoles/cm^2 per pulse). A correlation between extent of depolarization and increased Ca uptake therefore exists. These changes are illustrated in Figure 8.5. The resting value of $[Ca^{2+}]_i$ is approximately 0.1 μM, and an increase in $[Ca^{2+}]_i$ to at least 2 μM would be required to give a full contraction. It can easily be calculated that, as in the case of vertebrate skeletal muscle, the entering Ca falls very short of being able to produce this kind of a change in the value of $[Ca^{2+}]_i$.

The changes in $[Ca^{2+}]_i$ that accompany contraction have been measured optically using Ca-sensitive dyes. The relations between the depolarization produced during electrical stimulation, the tension produced by the contraction, and the value of $[Ca^{2+}]_i$ measured from the aequorin light signal are illustrated in Figure 8.6. Clearly, the Ca that entered the fiber during the depolarization cannot be responsible for the aequorin light response, as $[Ca^{2+}]_i$ rose to 0.4 μM in this experiment, and the known rate of Ca entry would increase $[Ca^{2+}]_i$ by, at most, one-tenth of this amount over the resting value. The light emission changes recorded with aequorin must reflect Ca release and reaccumulation within the elements of the SR. As with vertebrate

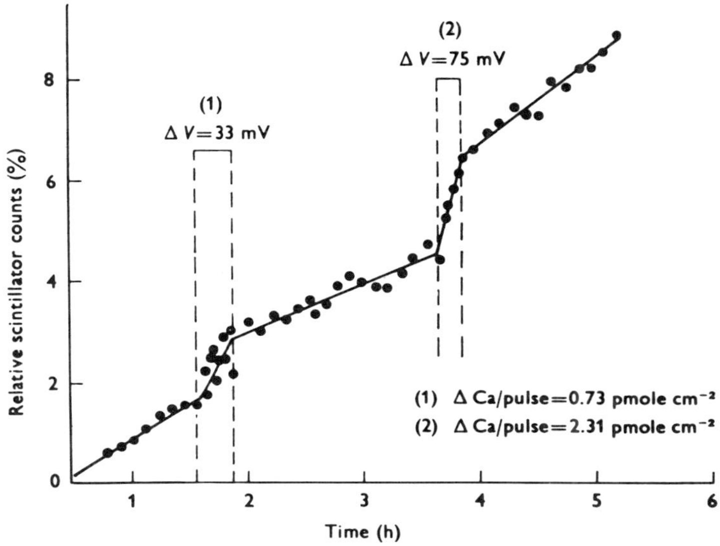

Figure 8.5. The effect of electrical stimulation upon the time course of the relative scintillation probe counts in a single *Balanus* muscle fiber. In the injected fiber, the final EGTA concentration after fiber dilution was 4.1 m*M*. Fiber diameter, 1.5 mm. Resting potential, −46 mV. From C. C. Ashley, P. C. Caldwell, A. K. Campbell, T. J. Lea, and D. G. Moisescu, Calcium Movements in Muscle, in *Symposium XXX of the Society for Experimental Biology*, C. Duncan, Ed., Cambridge University Press, Cambridge (1976), p. 397.

muscle fibers, the increased Ca influx during stimulation is of unknown significance, though entering Ca ions could serve as the trigger for a regenerative Ca release from the SR.[15]

INTRACELLULAR CALCIUM MOVEMENTS

Intrafibrillar Ca has been described as existing in two general compartments, the exchangeable and the nonexchangeable fractions. The exchangeable fraction is believed to be equivalent to the releasable Ca bound in the SR.[16] The nonexchangeable fraction is believed to consist largely of Ca bound to actin and myosin.[17] The transmembrane fluxes of Ca in intact frog muscle fibers discussed in the previous section can be summarized as amounting to about 0.1 pmole/cm^2·sec at rest. When the rate constant for the slow com-

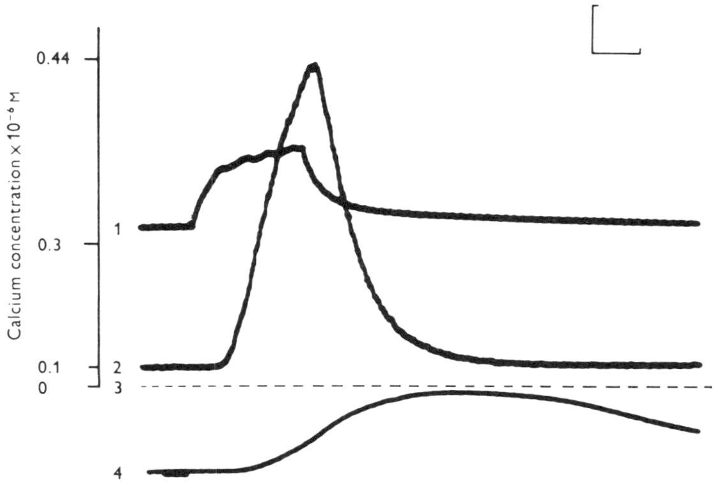

Figure 8.6. The effect of electrical stimulation upon the membrane response (trace 1), light emission from aequorin (trace 2), and isometric tension (trace 4) of a single *Balanus* muscle fiber injected with 0.2 ml of a saturated solution of aequorin in 100 m*M* TES buffer, pH 7.3. Trace 3 represents the baseline for $[Ca^{2+}]$ and light. The resting free calcium concentration is close to 0.1 μM. Calibration: vertical bar for trace 1, 20 mV; trace 4, 5 gwt. Horizontal bar = 100 msec. From C. C. Ashley, P. C. Caldwell, A. K. Campbell, T. J. Lea, and D. G. Moisescu, Calcium Movements in Muscle, in *Symposium XXX of the Society for Experimental Biology*, C. Duncan, Ed., Cambridge University Press, Cambridge (1976), p. 397.

ponent of Ca loss is applied to the Ca content of the exchangeable fraction, an efflux value of about 0.1 pmole/cm^2 · sec is obtained, so that in normal Ringer's solution, the Ca fluxes are approximately in balance. During stimulation, additional fluxes of Ca occur in both directions, about 0.20 pmole/cm^2 per impulse in magnitude. The time courses of the stimulated fluxes in intact fibers have not been resolved so that the relation between influx and efflux is not known. If the increased efflux is assumed to occur uniformly over a period corresponding to the contraction time of a fast twitch fiber, an extra efflux magnitude of 2–4 pmoles/cm^2 · sec is obtained. To analyze Ca movements in muscle further, use must be made of more refined techniques and preparations. As the surface membrane fluxes of Ca are inadequate to account for the rise in $[Ca^{2+}]_i$ that must precede contraction, somewhere within the muscle much larger Ca fluxes must occur.

One method that has been employed to resolve such intracellular Ca fluxes is that of the skinned muscle fiber.[18-21] In intact muscle fibers, the relatively low permeability of the surface membrane to Ca ions prevents the resolution of more rapid internal Ca fluxes by external means. In the skinned fiber preparation, a region of surface membrane is removed from intact frog muscle fibers in oil. These skinned segments respond to electrical stimulation and they also can be used in ^{45}Ca flux studies. In addition, the ends of the T-tubules apparently seal after skinning and permit ionic gradients to be set up across the tubular membranes. If fibers are kept at rest in a propionate medium, the introduction of bath chloride ions causes both Ca release from the SR and the development of tension, presumably via a depolarizing action on the tubular membranes. The uptake of Ca in such preparations can be measured by placing the skinned fibers in a bathing medium that contains a ^{45}Ca-EGTA buffer mixture of known specific activity. The efflux of Ca can then be analyzed by determining the ^{45}Ca lost to the bathing medium in given periods of time, the ^{45}Ca that remains in the myofibrillar space (MFS), and the ^{45}Ca that remains in the SR. The amount of isotope that remains in the MFS can be determined by using 1 m*M* EGTA to trap the ^{45}Ca in this compartment. The EGTA does not extract Ca from the SR. The SR Ca is obtained by extraction in Triton-X 100 and subsequent sequestration with 5 m*M* EGTA.[21]

Several facts have been learned from this preparation. If ionized Mg is low, a release of Ca from the SR can be obtained by adding bath Ca.[19] This Ca-induced Ca release may not occur at physiological values of $[Mg^{2+}]_i$, yet it may be significant in view of the suggestion that a small Ca movement may serve as the stimulus for a much greater Ca release. Also, the Cl-induced Ca release is Ca-dependent at low $[Mg^{2+}]$.

Much has been learned from the Cl-induced Ca release where the triggering depolarization spreads very rapidly over the entire fiber cross section as in normal excitation. The kinetics of ^{45}Ca washout can readily be obtained, as with intact fibers, by back-adding the counts lost to the solution and plotting data semilogarithmically versus time. When Ca is released by the SR, two processes occur simultaneously, namely, release and reaccumulation. These two processes can be separated by interrupting the normal response induced by Cl with addition of 1 m*M* EGTA at the peak of the response. This process removes MFS Ca ions, thus aborting the response. The amount of ^{45}Ca collected in this procedure can then be compared with the ^{45}Ca collected by 1 m*M* EGTA after the response has been allowed to go to completion. Accumulation of Ca by the SR will result in the collection of fewer ^{45}Ca counts in the MFS in the preparation in which the response has been allowed to proceed to completion. The amount of reaccumulation can then be obtained by subtraction of the completed MFS ^{45}Ca from the interrupted MFS ^{45}Ca.[21]

The main findings from this method are that the Cl stimulus produces a large increase in Ca efflux from the SR and that a large fraction of the released Ca is reaccumulated if EGTA is not present in the bath. During the first three seconds in the Cl bath, the tension rises steeply and about 16 percent of the fiber ^{45}Ca is released from the SR. Of this 16 percent, about 10 percent is still in the MFS and 6 percent has diffused out into the bathing collection solution. After seven seconds in Cl, about half of the released Ca is still in the fiber and subsequent loss of Ca to the wash solutions was small. About 70 percent of the ^{45}Ca that remained in the MFS after three seconds of Cl stimulation was never collected in the wash solutions and was therefore reaccumulated by the SR. It is of interest to examine the flux values themselves. The unstimulated or resting efflux from the SR corresponds to a concentration rate of change of 4–8 $\times$ 10^{-4} mM/sec per liter of fiber volume. The flux value that corresponds to this rate of concentration change depends on the membrane area involved. If restricted to the TC, the resting efflux would be 0.089 pmole/cm^2 · sec. The stimulated efflux from the SR is 100–200 times the resting efflux. As expressed above, these fluxes seem to be similar in magnitude to the surface membrane fluxes. This seems contrary to the earlier statement that, inside the fibers, much larger Ca ion fluxes are expected during stimulation. The answer lies with membrane areas. It is instructive to recalculate these fluxes on the hypothetical basis that they occur across the surface membrane of the fiber. The resting efflux from the SR would produce a flux of about 1 pmole/cm^2 · sec at the surface membrane. During stimulation, this flux would rise to at least 100 pmoles/cm^2 · sec. This means that this much flux would be required at the surface membrane to produce comparable rates of concentration change within the fibers. Comparison with actual surface membrane fluxes affirms that the surface fluxes are far too small to produce activation by themselves. An earlier calculation indicated that the actual surface efflux of Ca ions during stimulation might be about 2 pmoles/cm^2 · sec during the twitch. This is only 2 percent of the stimulated efflux from the SR in terms of the internal concentration changes produced, suggesting that, in the intact fiber, reaccumulation of released Ca is essentially complete. If, however, reaccumulation in the intact fiber is not perfect but is only 98–99 percent complete, the difference could easily be accounted for by the observed Ca efflux from the muscle fiber during stimulation. The purpose of the extra Ca influx into the fiber during stimulation might be, at least partially, to help preserve an overall steady state by replenishing the Ca lost to the exterior during contraction.

The skinned fiber method permits at least some quantitative estimates to be made of some of the rate processes illustrated in Figure 8.1. Another method that has proved useful is that of radioautography of ^{45}Ca.[4,22] This method indicates that the main site of release of Ca from the SR is located at the TC and the main sites of reaccumulation are along the longitudinal tubules with a slower return path to the TC within the longitudinal tubules (see Figure 8.1). By equilibrating fibers with known specific activities of ^{45}Ca, it is possible to estimate the rates of Ca movements between intracellular compartments. By this means, measurements of SR Ca movements in intact fibers are in essential agreement with the rates obtained on skinned fibers.

An interesting finding from the radioautographic studies is that, when fibers are first placed in a radioactive medium, the TC become quickly labeled with ^{45}Ca, whereas the rest of the SR is labeled at a much lower rate. The implication is that both the TC and the longitudinal tubules are not labeled from the sarcoplasm initially. As the TC are in close proximity to the transverse tubules, which, in the frog, are open to the exterior, it seems that the TC become quickly labeled with ^{45}Ca ions via the transverse tubular route. Conversely, when labeled fibers are transferred to an inactive solution for washout of ^{45}Ca, most of the ^{45}Ca collected during the first 10 minutes of washout comes from the TC, presumably via the transverse tubules. The permeability to Ca ions may thus be much higher at the specialized contacts between T-tubules and TC than at the surface membrane, at least under resting conditions.

REGULATION OF MUSCLE CALCIUM

The magnitude of the exchangeable Ca in muscle fibers varies between around 0.40 and 1.0 μmole/g fiber.[23] A good average value to bear in mind is 0.7 μmole/g. The value for the sarcoplasmic ionized Ca is about 10^{-7} M.[24] At rest, practically all of the exchangeable Ca is in the SR, maintained in that state by an ATP-dependent Ca pump.[25] At rest, relatively large turnovers of Ca occur within the fibers, and much smaller turnovers occur between the fiber and the external medium. During electrical stimulation, the rate of internal Ca exchange increases enormously (100–200 times), whereas only a modest increase in the external Ca exchange rate occurs across the sarcolemma. During the increased Ca recycling, which occurs within the fibers during stimulation, a shift of about one-third of the SR Ca, primarily from the TC, to the thin filament troponin molecules takes place. This Ca must be recycled to the TC during relaxation.

We now inquire as to the nature of the Ca flux balances that occur in resting muscle fibers. The internal resting Ca recycling can be regarded as mainly a pump-leak balance of fluxes between the SR and the sarcoplasm. The low value of $[Ca^{2+}]_i$ within the fibers is due energetically to the SR Ca pump activity. The very low value of $[Ca^{2+}]_i$ and the electrical negativity within the fibers combine to produce a large electrochemical driving force for Ca ions to enter the fibers from the exterior. The Ca equilibrium potential is given by

$$E_{Ca} = \frac{RT}{2F} \ln \frac{[Ca^{2+}]_o}{[Ca^{2+}]_i} \tag{88}$$

With a value of $[Ca^{2+}]_o$ of 2 mM, the calculated value for E_{Ca} is about 125 mV. For $E_m = -90$ mV, the total driving force is thus $125 + 90 = 215$ mV, and is in the direction favoring a large inward Ca ion leak, were P_{Ca} not low. The ^{45}Ca flux data of Bianchi and Shanes can be used to show that P_{Ca} is about equal to P_{Na} in frog muscle fibers.[1] On this basis, a passive electrodiffusional leakage rate of about 0.1 pmole/cm$^2 \cdot$sec would be expected for Ca ions entering the fibers. Though the SR could easily buffer such a Ca loading rate for a time, it could not do so indefinitely as its capacity is limited. There must, therefore, exist some means for the surface or transverse tubular membrane, or both, to extrude Ca ions from the interior to achieve a long-term steady state.

In general, cell membranes have at least two different mechanisms for extruding Ca ions from the cell interior. One mechanism is an ATP-dependent Ca pump similar to the Ca pump in the SR. The other mechanism is a Na:Ca counter-transport process in which the downhill energy driving Na ions into the cell is coupled to the outward extrusion of Ca ions. Evidence for a Ca pump driven by ATP has been obtained in squid giant axons.[26] Evidence for Na:Ca counter transport has also been obtained in squid giant axons.[27-29] The operation of a Na:Ca counter-transport process has been demonstrated in frog heart ventricular muscle.[30-32] The Na:Ca counter-transport process can manifest itself in several ways. As Na ions entering the cell extrude Ca ions, immersion in Na-free media can result in an increase in internal Ca levels. An $[Na]_o$-dependent Ca efflux can occur such that removal of external Na results in a diminution in Ca efflux. In a lithium medium, a $[Ca^{2+}]_o$-dependent Na efflux occurs that is inhibited by external Na.[27] Under such conditions, an increased Ca influx is associated with the Ca_o-dependent Na efflux. A number of these effects have been observed in giant striated muscle fibers from the barnacle. When the Ca-sensitive photo-protein aequorin is used to monitor $[Ca^{2+}]_i$, replacement of external Na by Li, choline, sucrose, or tris results in a rapid increase in the level of light emission.[33] The response is dependent on the presence of external Ca^{2+} and indicates that an increase in $[Ca^{2+}]_i$ within the fibers occurs due to either an increased rate of Ca entry or a decreased rate of Ca exit from the fibers, or to both. The results are those expected if the Na gradient is responsible for a portion of Ca extrusion in barnacle muscle fibers. Results in which an internal scintillator probe is used to measure ^{45}Ca influx in barnacle fibers show that Ca influx is increased in a Na-free medium when Li is used to replace Na.[2] In barnacle muscle fibers microinjected with ^{45}Ca, the efflux of Ca is found to depend on both external Ca and Na levels. Removal of external Ca results in a drop in Ca efflux and the subsequent removal of external Na results in a further decline.[33,33a] The effects on Ca efflux were observed whether Li, choline, or sucrose was used as a Na replacement. The results of the Ca flux studies are summarized in Figures 8.7 and 8.8. All of these data are consistent with the Na-gradient theory for Ca regulation.

It is worthwhile to examine the feasibility of the Na-gradient theory for muscle fibers quantitatively. If the sole source of energy for Ca transport is the Na electrochemical gradient, the thermodynamic relation that must hold at equilibrium where the Na and Ca gradients just balance one another is, from Mullins,[34]

$$qF(E_{\mathrm{Na}} - E_m) - 2F(E_{\mathrm{Ca}} - E_m) = 0 \tag{89}$$

where q is the number of Na ions that are transported per Ca ion. This expression can be readily transformed to give, for the Ca ratio,

$$\frac{[\mathrm{Ca}^{2+}]_o}{[\mathrm{Ca}^{2+}]_i} = \left(\frac{[\mathrm{Na}^+]_o}{[\mathrm{Na}^+]_i}\right)^q e^{(2-q)E_m F/RT} \tag{90}$$

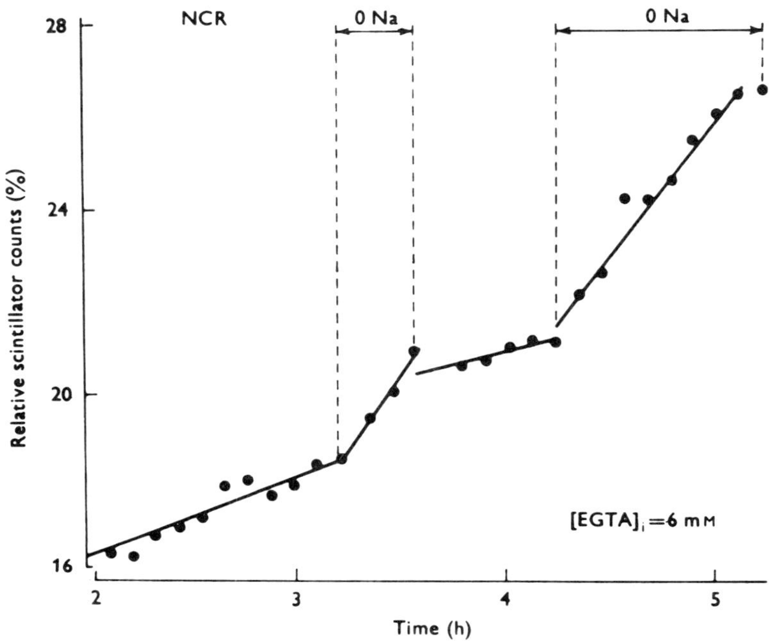

Figure 8.7. The effect of Na replacement, 0 Na (Li), upon the time course of the relative scintillator counts for a single *Balanus* muscle fiber. NCR: normal crab Ringer's solution. Fiber diameter, 1.2 mm; resting potential, −43.0 mV. From C. C. Ashley, P. C. Caldwell, A. K. Campbell, T. J. Lea, and D. G. Moisescu, Calcium Movements in Muscle, in *Symposium XXX of the Society for Experimental Biology*, C. Duncan, Ed., Cambridge University Press, Cambridge (1976), p. 397.

Applied to barnacle muscle fibers where E_m is about -50 mV, the Ca ratio is calculated from equation (90) to equal about 2×10^5 when $q = 4$. For $[Ca^{2+}]_o = 10$ mM, $[Ca^{2+}]_i$ would equal 0.05 μM, a value well within the range of measured values. For frog skeletal muscle fibers with a membrane potential of -90 mV, there is enough energy in the Na gradient to account for the Ca distribution between the sarcoplasm and exterior even if only three Na ions are transported per Ca ion.

As much ion flux work has been done on frog muscle fibers, it is natural to inquire as to the presence of the Na : Ca transport mechanism in this preparation. The fact that the mechanism is theoretically feasible on thermodynamic grounds does not constitute evidence for its presence. That Na : Ca transport is found in other excitable fibers, however, makes it reasonable to suppose

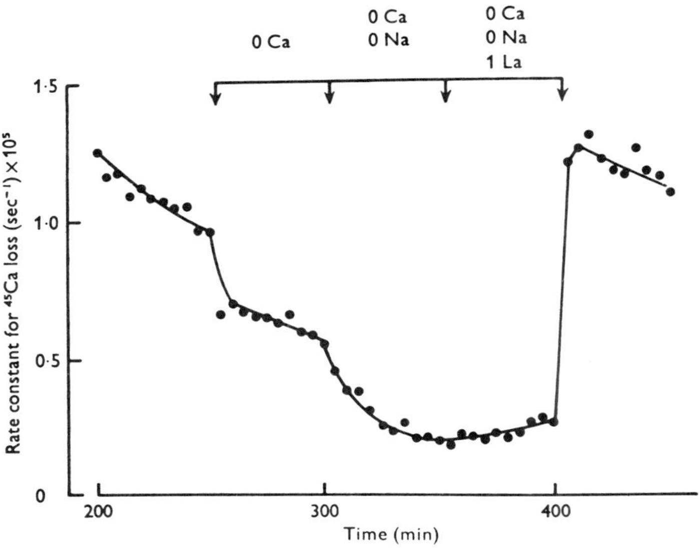

Figure 8.8. The effect of replacing external Ca and Na, and the addition of 1 m*M* La^{3+} on the efflux of ^{45}Ca from a single *Balanus* muscle fiber. Resting potential, −48 mV. From C. C. Ashley, J. C. Ellory, and K. Hainaut, *J. Physiol.* **242,** 255 (1974).

that the mechanism is present in vertebrate skeletal muscle as well. A search for hard scientific evidence reveals that data in support of the Na:Ca mechanism in vertebrate skeletal muscle are scant and somewhat conflicting. A relation between total muscle Ca and muscle Na and K levels in frog muscle has been reported.[35] Muscles with elevated Na content and depleted K levels showed heightened levels of total analytical Ca. Though these authors interpret their work on the basis of a Ca-for-K exchange process in the fibers, the results could equally well be explained on the basis of the Na-gradient theory. As muscles gain Na and lose K, the membrane potential declines unless electrogenic Na pumping keeps pace to maintain the potential near normal values. It is easy to visualize how the magnitude of the Na gradient could decline under such conditions, giving rise to elevated values of $[Ca^{2+}]_i$ that would, in turn, increase the total Ca level by promoting increased Ca uptake by the SR. Using the Na_i and Ca_i levels cited, and assuming that $[Ca^{2+}]_i$ has risen in proportion to total Ca_i, one calculates from the data of Cosmos and Harris[35] that more than enough decline in the Na gradient has occurred to account for the elevated Ca levels, even if an equilibrium has not been reached in the fibers. One could state, therefore, that the data are consistent with an Na-gradient mechanism for Ca regulation.

The strong relation between Ca content and Na and K contents cited above has not been confirmed in other laboratories, however. The results of one study in which Na and K contents of the muscles were varied by storage in K-free media for different times are summarized in Table 8.1.[36] The total analytical Ca content of the muscles does not vary significantly when $[Na]_i$ varies over a sevenfold range. The conclusion from this study is that the total internal stores of Ca ions in frog muscle do not depend in any crucial way on the internal ionic state of the fibers with respect to Na and K ions. Any discrepancies in this area will have to await further experimentation for clarification.

There is also some evidence in frog skeletal muscle that Ca fluxes are dependent on the external Na concentration as expected from the Na-gradient theory. When ^{89}Sr was used as a tracer for Ca ions, reduction in $[Na^+]_o$ resulted in increases in the rate of ^{89}Sr uptake. The uptake rate of ^{89}Sr was maximal in an Na-free medium.[35] Sucrose was used to replace Na in this study and the uptake rate for ^{89}Sr was increased by three- to fourfold in an Na-free medium. Again, these data support the Na-gradient model for Ca transport in vertebrate skeletal muscle. In another study using ^{45}Ca, Na-free solutions had no great effect on the rate of Ca uptake into the slowly exchanging compartment of the muscle fibers. After one hour of uptake in Na, the slow compartment had exchanged 0.071 ± 0.03 μmole/g of Ca.[36] Also, muscles stored in Li Ringer's solution for several hours did not show any significant net gain of total Ca ions over control muscles stored in Na Ringer's

Table 8.1. Calcium Contents of Muscles at Different Sodium and Potassium Contents in μmoles/g ± SD

Ca_i	Na_i	K_i
1.65 ± 0.09	5.06 ± 0.18	90.5 ± 8.0
1.85 ± 0.24	19.1 ± 5.0	75.6 ± 7.4
1.78 ± 0.04	37.2 ± 3.0	50.6 ± 3.0

Source: From R. A. Sjodin and S. Gold, unpublished results (1978).

solution for the same period of time. Though the effect of Li ion substitution on ^{45}Ca uptake was small in these studies, there was quite a significant influence of Li ion substitution on ^{45}Ca washout from the fast compartment or compartments. The efflux of ^{45}Ca from the more rapidly exchanging compartment is much higher in Na media than in Li Ringer's solution.[36] These results are illustrated in Figure 8.9. For one muscle, ^{45}Ca washout was measured in Na Ringer's solution initially, while ^{45}Ca washout from the paired muscle was measured initially in Li Ringer's solution. During the early phase of ^{45}Ca washout, much more ^{45}Ca was lost from the muscle in the Na medium than from the muscle in the Li medium. When the solutions were subsequently changed from Li to Na and from Na to Li, respectively, there was no change in ^{45}Ca efflux from the slowly exchanging compartment. Clearly, it is the rapidly moving Ca that is sensitive to external Na and not the slowly moving Ca. In view of the fact that muscles placed in ^{45}Ca solutions first show labeling in the TC and first lose ^{45}Ca from the TC when placed in a nonradioactive solution,[4,22] it is likely that the Ca fluxes in the TC are the ones sensitive to external Na. Further work is necessary to clarify the mechanism and role of Na-sensitive Ca fluxes in vertebrate skeletal muscle.

SODIUM:CALCIUM TRANSPORT AND THE RESTING MEMBRANE POTENTIAL

If more than two Na ions are transported inwardly per Ca ion transported outwardly during Na:Ca transport, as is the case for muscle fibers, the process will be electrogenic, and a contribution to the resting membrane potential will occur. This contribution is, of course, in addition to the electrogenic contribution of the Na:K pump. Previously developed equations for E_m do not take Na:Ca transport into account. If the contribution to E_m of Na:Ca transport is significant, equation (53) will not be valid. An equation taking all previously discussed processes into account has been developed.[37] With Ca ion permeability taken into account with no Na:Ca transport or electrogenic Na:K transport present, the equation for the membrane potential becomes

$$E_m = \frac{RT}{F} \ln \frac{4P'_{Ca}[Ca]_o + P_K[K]_o + P_{Na}[Na]_o + P_{Cl}[Cl]_i}{4P''_{Ca}[Ca]_i + P_K[K]_i + P_{Na}[Na]_i + P_{Cl}[Cl]_o} \tag{91}$$

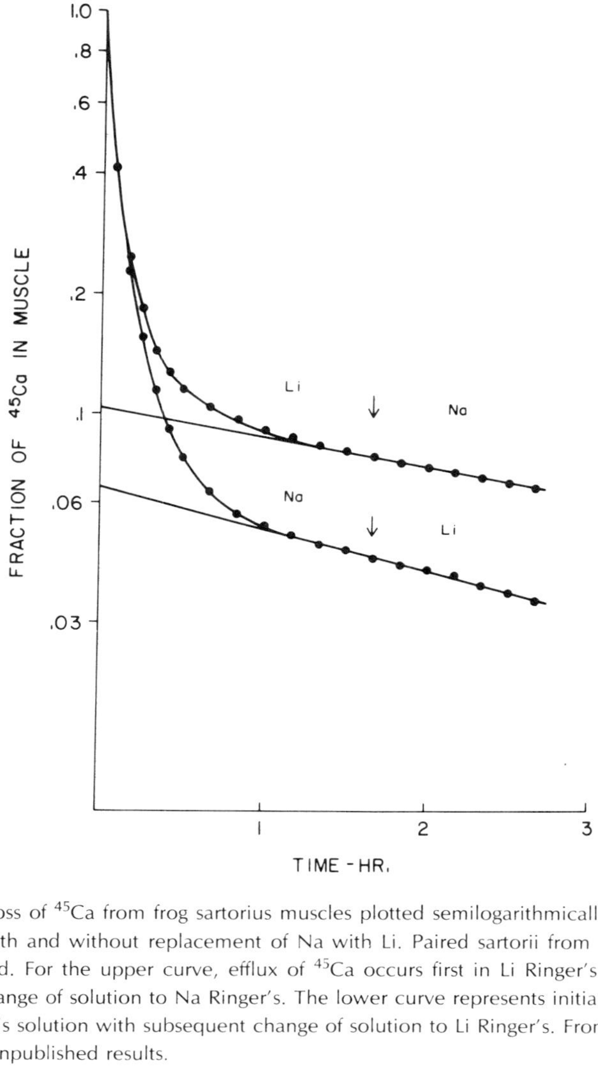

Figure 8.9. Loss of ^{45}Ca from frog sartorius muscles plotted semilogarithmically against time in solutions with and without replacement of Na with Li. Paired sartorii from the same frog were employed. For the upper curve, efflux of ^{45}Ca occurs first in Li Ringer's solution with subsequent change of solution to Na Ringer's. The lower curve represents initial efflux in normal Na Ringer's solution with subsequent change of solution to Li Ringer's. From R. A. Sjodin and S. Gold, unpublished results.

where

$$P'_{\mathrm{Ca}} = P_{\mathrm{Ca}} \left(e^{E_m F/RT} + 1\right)^{-1} \tag{92}$$

and

$$P''_{\mathrm{Ca}} = P'_{\mathrm{Ca}}\, e^{E_m F/RT} \tag{93}$$

To take Na:Ca counter transport into account, it is necessary, in addition to terms described previously, to define the flux of Na by Na:Ca transport as t_{Na} and the flux of Ca by Na:Ca transport as t_{Ca}. For the steady state to hold for each cation, the following relations must be obeyed:

$$j_{\mathrm{Ca}} = -t_{\mathrm{Ca}} \tag{94}$$

$$j_{\mathrm{Na}} + t_{\mathrm{Na}} = -m_{\mathrm{Na}} \tag{95}$$

$$j_{\mathrm{K}} = -m_{\mathrm{K}} \tag{96}$$

where the passive leaks, j, for each ion are given by equation (9) and where m_{Na} and m_{K} are fluxes due to the Na:K pump. Using equation (50) and introducing the stoichiometric ratio, q, for Na:Ca transport,

$$q = -\frac{t_{\mathrm{Na}}}{t_{\mathrm{Ca}}} \tag{97}$$

the following equation is obtained for the membrane potential using equation (9) for each ion:

$$E_m = \frac{RT}{F} \ln \frac{2qP'_{\mathrm{Ca}}[\mathrm{Ca}]_o + rP_{\mathrm{K}}[\mathrm{K}]_o + P_{\mathrm{Na}}[\mathrm{Na}]_o}{2qP''_{\mathrm{Ca}}[\mathrm{Ca}]_i + rP_{\mathrm{K}}[\mathrm{K}]_i + P_{\mathrm{Na}}[\mathrm{Na}]_i} \tag{98}$$

A knowledge of permeability coefficients, ionic concentrations, and the stoichiometric coefficients r and q enable calculation of the membrane potential with both electrogenic Na:K transport and Na:Ca transport present. When a steady state is not present, equations similar to (67) and (68) can be written.

$$f_{\mathrm{Na}}(j_{\mathrm{Na}} + t_{\mathrm{Na}}) = -m_{\mathrm{Na}} \tag{99}$$

$$f_{\mathrm{K}} j_{\mathrm{K}} = -m_{\mathrm{K}} \tag{100}$$

and

$$f_{\mathrm{Ca}} j_{\mathrm{Ca}} = -t_{\mathrm{Ca}} \tag{101}$$

The equation for E_m, when ions are not in a steady state, is

$$E_m = \frac{RT}{F} \ln \frac{2qf_{\text{Ca}}P'_{\text{Ca}}[\text{Ca}]_o + r(f_{\text{K}}/f_{\text{Na}})P_{\text{K}}[\text{K}]_o + P_{\text{Na}}[\text{Na}]_o}{2qf_{\text{Ca}}P''_{\text{Ca}}[\text{Ca}]_i + r(f_{\text{K}}/f_{\text{Na}})P_{\text{K}}[\text{K}]_i + P_{\text{Na}}[\text{Na}]_i} \qquad (102)$$

When there is no electrogenic Ca transport, that is, either no Na:Ca transport or $q = 2$, and P_{Ca} is negligible for existing [Ca] values, equation (102) reduces to equation (70).

Normally, an inward stream of Na ions extrudes Ca ions. If more than two Na ions are transported per Ca ion, the inward charge movement per one Na:Ca transport cycle will result in membrane depolarization. The larger the value of P_{Ca}, the larger will be the depolarization for any particular ionic state, because the rate of Na:Ca transport will have to be larger to maintain that state. The extent of depolarization by electrogenic Na:Ca transport for various relative transport rates and relative permeability ratios is illustrated in Figure 8.10. The uppermost curve corresponds to the relative permeability ratios holding in frog muscle. The relatively small effect of Na:Ca transport under these conditions no doubt explains why it has been possible to successfully describe the membrane potential for these fibers without taking Na:Ca transport into account. Another reason is that q is probably smaller for frog muscle than the value of 4 used in the calculations. If $q = 3$ for frog muscle, additional depolarization due to Na:Ca transport will be even less than that depicted in Figure 8.10. For the P_{Ca} values in frog skeletal muscle, the terms involving Ca ions would be negligible for normal values of $[\text{Ca}^{2+}]$. In this case, equation (91) reduces to equation (39), which has been very successful in describing E_m for frog muscle. In summary, Na:Ca transport is not likely to make a significant contribution to the resting membrane potential for frog muscle fibers under normal conditions.

In giant muscle fibers from the barnacle, Na:Ca transport seems to be definitely present as evidenced by flux studies. Thermodynamic calculations indicate that q would have to have a value of around 4 for the process to be effective in Ca regulation. One would expect, therefore, a contribution of Na:Ca transport to the membrane potential of barnacle muscle fibers. A lack of quantitative knowledge concerning the value of P_{Ca} makes it impossible at present to predict the degree of depolarization expected in this preparation. However, at elevated values of $[\text{Ca}^{2+}]_i$ an Na_o-dependent depolarization of around 3 mV has been observed, confirming the theoretical prediction that Na:Ca transport is electrogenic.[38]

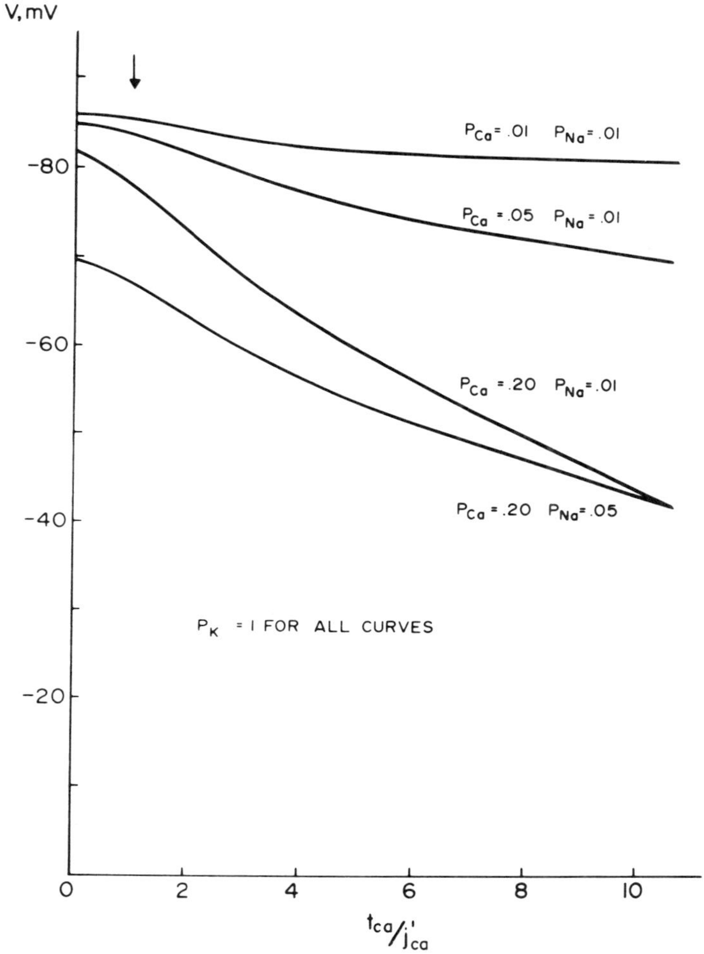

Figure 8.10. Contribution of Na:Ca transport to the resting potential for different permeability ratios $P_K:P_{Na}:P_{Ca}$. The membrane potential is calculated from equation (102) against increasing ratios of the Na:Ca transport rate to the inward Ca leakage rate. The value of $r = 1.5$, $q = 4.0$, $[Na]_o = 150$ mM, $[K]_o = 5$ mM, $[Ca]_o = 2$ mM, $[Na]_i = 10$ mM, and $[K]_i = 150$ mM. From R. A. Sjodin, *J. Gen. Physiol.* **76,** 99 (1980).

REFERENCES

1. C. P. Bianchi and A. M. Shanes, *J. Gen. Physiol.* **42,** 803 (1959).
2. C. C. Ashley, P. C. Caldwell, A. K. Campbell, T. J. Lea, and D. G. Moisescu, Calcium Movements in Muscle, in *Symposium XXX of the Society for Experimental Biology,* C. Duncan, Ed., Cambridge University Press, Cambridge (1976), p. 397.
3. C. C. Ashley and E. B. Ridgeway, *J. Physiol.* **209,** 105 (1970).
4. S. Winegrad, *J. Gen. Physiol.* **51,** 65 (1968).
5. D. L. Gilbert and W. O. Fenn, *J. Gen. Physiol.* **40,** 393 (1957).
6. A. M. Shanes and C. P. Bianchi, *J. Gen. Physiol.* **42,** 1123 (1959).
7. B. A. Curtis, *J. Gen. Physiol.* **50,** 255 (1966).
8. A. L. Hodgkin and P. Horowicz, *J. Physiol.* **153,** 386 (1960).
9. G. B. Frank, *J. Physiol.* **151,** 518 (1960).
10. B. A. Curtis, *J. Gen. Physiol.* **47,** 953 (1964).
11. C. M. Armstrong, F. M. Bezanilla, and P. Horowicz, *Biochim. Biophys. Acta* **267,** 605 (1972).
12. E. Stefani and D. J. Chiarandini, *Pflügers Arch.* **343,** 143 (1973).
13. H. Ch. Lüttgau and W. Spiecker, *J. Physiol.* **296,** 411 (1979).
14. C. P. Bianchi and A. M. Shanes, *J. Cell Comp. Physiol.* **56,** 67 (1960).
15. L. E. Ford and R. J. Podolsky, *Science* **167,** 58 (1970).
16. S. Ebashi and M. Endo, *Prog. Biophys. Mol. Biol.* **18,** 123 (1968).
17. W. Hasselbach, *Prog. Biophys. Mol. Biol.* **14,** 167 (1964).
18. L. E. Ford and R. J. Podolsky, *J. Physiol.* **223,** 1 (1972).
19. L. E. Ford and R. J. Podolsky, *J. Physiol.* **223,** 21 (1972).
20. E. W. Stephenson and R. J. Podolsky, *J. Gen. Physiol.* **69,** 17 (1977).
21. E. W. Stephenson, *J. Gen. Physiol.* **71,** 411 (1978).
22. S. Winegrad, *J. Gen. Physiol.* **55,** 77 (1970).
23. C. P. Bianchi, *Cell Calcium,* Appleton-Century-Crofts, New York; Butterworth, and Co. Ltd., London (1968).
24. M. Endo, *Physiol. Rev.* **57,** 71 (1977).
25. A. P. Carvalho and B. Leo, *J. Gen. Physiol.* **50,** 1327 (1967).
26. R. DiPolo, *J. Gen. Physiol.* **69,** 795 (1977).

27. P. F. Baker, M. P. Blaustein, A. L. Hodgkin, and R. A. Steinhardt, *J. Physiol.* **200,** 431 (1969).

28. M. P. Blaustein, J. M. Russell, and P. DeWeer, *J. Supramol. Struct.* **2,** 558 (1974).

29. J. Requena, R. DiPolo, F. J. Brinley, Jr., and L. J. Mullins, *J. Gen. Physiol.* **70,** 329 (1977).

30. R. Niedergerke and E. J. Harris, *Nature* (Lond.) **179,** 1069 (1957).

31. H. C. Lüttgau and R. Niedergerke, *J. Physiol.* **143,** 486 (1958).

32. R. Niedergerke, *J. Physiol.* **167,** 515 (1963).

33. C. C. Ashley, J. C. Ellory, and K. Hainaut, *J. Physiol.* **242,** 255 (1974).

33a J. M. Russell and M. P. Blaustein, *J. Gen. Physiol.* **63,** 144 (1974).

34. L. J. Mullins, *J. Gen. Physiol.* **70,** 681 (1977).

35. E. Cosmos and E. J. Harris, *J. Gen. Physiol.* **44,** 1121 (1961).

36. R. A. Sjodin and S. Gold, unpublished results, (1978).

37. R. A. Sjodin, *J. Gen. Physiol.* **76,** 99 (1980).

38. M. T. Nelson and M. P. Blaustein, *Biophys. J.* **33,** 61a (1981).

Index